W0245735

Schriftenreihe
der Wissenschaftlichen Landesakademie
für Niederösterreich

Sabine Stifter, Jadran Lenarčič (eds.)

Advances in Robot Kinematics

1991. 208 figures. XIV, 484 pages.
Soft cover DM 148,-, öS 1036,-
ISBN 3-211-82302-6

Prices are subject to change without notice

This volume includes a selection of papers presented at the second workshop on Robot Kinematics held in Linz, September 10–12, 1990.

The papers present new results and overviews on various aspects of robot kinematics such as modelling and computation, analysis and design, motion planning and control, inverse kinematics calculations, kinematic redundancy, and parallel mechanisms. Special emphasis was put on the investigation of symbolic computation techniques for problems in robot kinematics.

Springer-Verlag Wien GmbH

Preface

Austria has a long tradition in interdisciplinary research in IFAC. For example in 1983 the first IFAC Workshop on "Supplemental Ways for Improving International Stability (SWIIS)" was organized in Laxenburg near Vienna. On this occasion the IFAC Working Group on "SWIIS" was founded. During the last IFAC Congress in Tallinn the Committee on Social Effects of Automation installed a new Working Group on *"Cultural Aspects of Automation"* and the Austrian NMO suggested to have the first Workshop on this topic in Austria.

According to one of the main goals of the Scientific Academy of Lower Austria in Krems this Workshop took place there. The organizer, the "Austrian Center for Productivity and Efficiency" as the Austrian IFAC-NMO and the Department of "Systems Engineering and Automation" of the Academy of Lower Austria welcomed participants from different research disciplines, like control engineering, systems engineering, sociology, art, philosophy, and politics. The international programme committee had to select papers of high scientific level.

These papers served only for initiating intensive discussions. Main topics of these discussions were technology design, automation software and culture, social conditions, education, computer and art, design of man-machine-systems, CIM and culture as well as appropriate methods for interdisciplinary research. Approximately 50 % of the participants were from Eastern Europe. Therefore special emphasis was given to the influence of automation at different cultures, especially Eastern and Western cultures. As a first result from the Workshop one of the next tasks of this new IFAC-Working Group should be the design, development, and construction on automation devices taking under consideration the cultural aspects. This first Workshop has fullfilled the main goal for scientists of different research disciplines to define a common vocabulary working together on research projects.

In this Proceedings Volume the scientifc contributions to this Workshop are collected. The authors had to finish their manuscripts in camera-ready form. Therefore the editors aren´t responsible for the contents of each contribution as well as the grammar.

The editors thank the contributors for their efforts to offer high-quality material, the publishing company and A. Weichselbaum, N. Stanek and A. Binder for retyping and editing some papers.

Krems, December 1991

P. Kopacek
NOC-Chairman

J. Forslin
IPC-Chairman

Contents

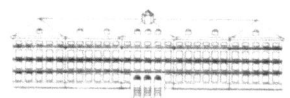

J. Forslin and P. Kopacek (eds.)

Cultural Aspects of Automation

Proceedings of the 1st IFAC Workshop on Cultural Aspects of Automation, October 1991, Krems, Austria

Springer-Verlag Wien GmbH

J. Forslin
University of Bergen, Bergen, Norway

P. Kopacek
Scientific Academy of Lower Austria, Krems, Austria

With 21 Figures

ISBN 978-3-211-82362-0   ISBN 978-3-7091-9220-7 (eBook)
DOI 10.1007/978-3-7091-9220-7

Opening of the Workshop

N. Roszenich
Director General
Ministry for Science and Research
Vienna, Austria

Ladies and Gentlemen,

let me welcome you to this meeting here in Austria on behalf of the Ministry for Science and Research.

We are happy to have you here and I am looking forward to fruitful discussions in the days to come.

The presentation during this conference will deal with the interplay between recent developments in automation and the culture and social framework, with special emphasis on the approaches both in the East and the West.

Mechanical devices performing simple tasks, such as the "Jacquard" - loom have been in wider use already since last century. The relationship between automation on the one side and cultural and social values on the other side has occupied human imagination especially since the appearances of "Robots" in movies and science-fiction stories.

The distinguished writer Stanislaw Lem from Poland wrote in his collection of short stories, "Bajki robotów" (robot's fairy tales), published in Krakow in 1964, about the attempts of two scientics to construct powerful armies out of individual soldiers, by equipping each of them with electrical connections to his neighbours.
The emperors of two fighting kingdoms did hope that this linkage between the individual minds of the soldiers would lead to the formation of a disciplined and united military mind.

But a quite different development evolved: by linking in ever greater number, the conscience changed and crossed the border towards cosmic dimensions. And the cosmos is - according to Lem - devoid of military ideas: the two armies were on the outside still displaying their steel, armour and the deadly weapons, but inside they contained two oceans full of happiness, of all- embracing goodwill and perfect reasons.

And thus, the two emperors, aghast with rage and shame, could not help but watch "the two armies strolling about, linking hands, collecting flowers on the field of the great battle that did not take place."

A quarter of a century has passed since Stanislaw Lem wrote this fairy tale about human values and the impact of automation.

You may recall that at the time the first transistors were used in products. Computers were noisly roomsized boxes; strange looking "robots" were shown in news-reels. The process of industry and technology promised a bright future to everyone....

Looking back now, we realize how much the perception of our common world has changed since: technology can no longer bring the solution to our problems and fullfill all our needs.
Now we learn about the importance of a "substainable development": the concept of a society which is in broad balance with the availability of resources.

Facing industrial pollution, dwilling resources and global warming, we see plainly that terms such as "growth" and "development" can be quite misleading at times.

The topics of the contributions to this conference here bear witness to these profound changes. They address "man-machine-systems" and the question of "automation and the human search for freedom". They will present research on "cultural aspects" and discuss "human-centered technologies".

I am confident that the two days ahead will be very stimulating. Research in automation and expert systems surely has not only been the subject of science-fiction novels, it has attracted outstanding sciences as well.

As the topics of the following presentations show us, the interest of these researchers extends well into areas of vital and general interest to all of us.

Again, let me wish you a pleasant stay at this conference.

From Fragmentation to Integration
Summary of an Introduction to a Workshop on Cultural Aspects of Automation

J. Forslin
University of Bergen
Bergen, Norway

Today there is a heated debate on what are the future forms of work and how technology should be best utilized to meet both economic and social ends. In parallel there is sharpened global competition in business, where technology in itself is less seen as the prime weapon. It is rather the way it is being utilized, which depends on human resources: skills, creativity, values, commitment etc. that gives the edge. Such factors have a cultural background and we thus have to start to look at culture as a competitive advantage - or for that matter even disadvantage. How to best make us of a culture's strong side is now becoming a profound issue for research, where technical and social expertise have to meet.

Technology, however primitive, has always been accompanying man as a tool for survival. From having emerged as a hunter gatherer to become a farmer some 12 000 years ago man only recently became an industrial worker. This development has not been uniformly positive and has been full of crises and suffering. There has also been increasing room for improvement though, as the accumulation of knowledge has progressed.

Accumulation of knowledge has been accompanied by accumulation of capital. Out of these processes industry was born. Industry as a technology was not primarily based on technical advances, but rather on a new way of thinking and new principles. These principles can be boiled down to that of fragmentation. First capital, labour and know-how, once united in one producer under craft production, were split and gradually reflected in a new social structure. Further, the work process itself was strongly divided along a horizontal and vertical dimension, increasing the productivity of man and facilitating the application of technical inventions - work organization and technical progress went hand in hand.

The system can be said to rest on four pillars:

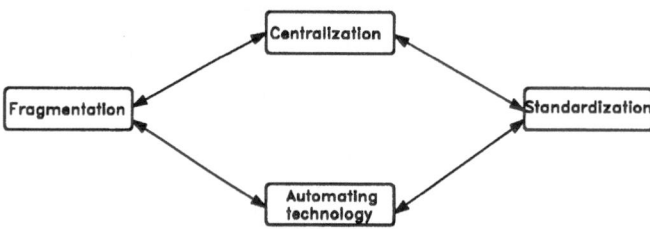

This highly productive concept, however, also contained social tensions and sources of serious social conflicts. Also with increasing material welfare, values with regard to work have changed and aspirations increased, which have profoundly challenged the predominante way of organizing the production process. Also the business environment has changed into a highly turbulent and complex world. The once so successful industrial system is now seen as less and less able to handle the new situation from either social or economic point of view.

The system backfield some 25 years ago, first socially when people at the shop floor in principle refused to go to work anymore. That problem has been the prime force for research efforts to create more attractive ways of arranging industrial work processes. In particular in the northern Europe the solution - from Kalmar to Uddevalla to use the Swedisch Volvo- approach as an illustration - have basicly had a sociotechnical orientation, a theory that can be seen as a forerunner to the new technology to come.

Since then information technology has emerged as a revolutionary force. When the industrial system todays stands out as antiquated one should search for the ultimate reason in the advancing information technology with all its corollaries. The problem of today is slowness in adapting to the new conditions and prevailing obsolete views on organization and human resources. In the emerging Info Tech-society four new pillars can be identified derived from the inverse operation of fragmentation - integration.

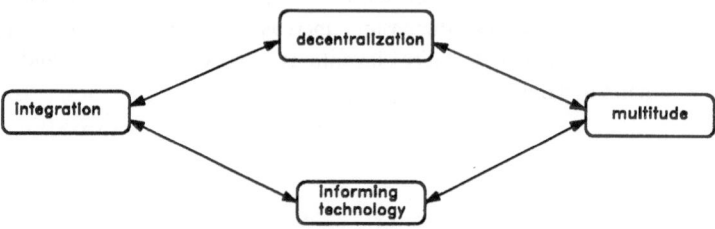

Based on these elements the organizations of the future will look very different from today. There will also be less of standardized solutions, rather in order to optimize in a situation of complex demands each enterprise shall find its own unique way of operation. As a response to an ever ongoing process of change the organizations ability for renewal will be as important as production - the learning organization.

The philosophy behind automation was to get rid of the human factor. Information technology in stated strengthens and expands human faculties and makes man indispensable. The move from standardization means, that not only each enterprise, but also each country has to find its unique ways of dealing with the new situation. The solutions at enterprise level have to reflect the unique cultural context and bring out the best of it.

What then is culture? There are virtually thousands of definitions of this concept. Here is offered a specific view as a frame of references: culture is the element by which wholeness is created in a society. Culture connects the technical, the social and the spiritual levels by creating a unique blend of values, mores, traditions, interpretations etc. It has a specific history meaning, but also contains sediments from the past as well as out-looks for the future. The social effects of a technical innovation can only be understood against the background of the specific cultural context and it is the opportunity of this workshop to contribute to a more complete understanding of the dynamics of technical advances.

Background and Interfaces of Automation in Finland

A.J. Niemi
Laboratory of Control Engineering
Helsinki University of Technology
Helsinki, Finland

ABSTRACT

Basic automation in Finnish industry proceeded parallelly to mechanisation in primary production and to the large social change caused by the latter. Government and industry support automation, while the public and trade unions are neutral or in favour of it. While it is expanded in industry, a scientific backing is seldom sought for.

1. Mechanisation and Automation in Finland up to Now

Automation has in Finland proceeded as a natural step of industrial development. It has been preceded by the mechanisation of operations which was the essential content of the first industrial revolution and consisted of transfer from muscle power to machine power in the industries. This previous step happened with a considerable delay in some industries, and especially so in agriculture and forestry which earlier employed a majority of the working power in Finland.

Mechanisation of forestry started only after the second world war, with transfer from the traditional saw to the chain saw and continued later with transfer to harvester. At the same time, the floating of logs was largely changed to road transport with trucks, and horses in the woods and similarly on the fields were substituted by tractors. These processes effected a decrease of the labor force employed in the primary production from 46% (of a total labor force of 2.000.000) in 1950 or 35% in 1961 to 12,6% (of 2.200.000) in 1980, or with 630 000 from 1950 to 1980. This decrease is continuing still today, although at a lower speed.

While the unemployment was low up to 1976, the released part and the natural increase of the labor force were absorbed by the industry and construction, and the service industries sectors. Despite the emigration, especially to near-by Sweden which at that time offered plenty of opportunities in the manufacturing industry, the change from the primary production to the industry and construction (more than 34% in 1979) and service industries (more than 52% in 1979) was in Finland one of the fastest in the industrialized world. This development happened parallelly to an increase of the total production which in 1974 was more than three times that in 1950, or to an average annual increase of the GNP with more than 5% which exceeded the corresponding number of the OECD countries with 1%.

The typical industry in Finland are the process industries which have been machine-powered from their origination during the last century. Accordingly, they have never experienced the first industrial revolution. The second one, i.e. the automatization of man's decision and control actions by means of instrumentation, started in early 1950's. A major part of such tasks at the production line was automatized in ten years in the pulp and paper industries and power plants, and during the following decade in the other process industries. Thus the automation proceeded simultaneously with the mechanisation of the primary production. Available statistics do not distinguish its effect on the labor force from those of other technologial developments or of the industrial expansion, but certainly they were much smaller than those of the described mechanisation.

Mechanical mass production of the type of car idustry lacks almost totally in Finland, and therefore the hard or Detroit type automation has never been really applied. To some degree, a similar production can be found in saw-mills in which machines do the work but transfers of work pieces have been manually executed or aided. These operations have been mechanised or automatized from 1960's onwards which has decreased the need of labor per produced unit. However, a direct substitution of men with machines or instruments has usually not taken place, because the level of automation in saw-mills is heavily correlated with the production capacity of plant. That is to say: big and highly automated plants have been built and smaller, less automated plants have elsewhere laid down their operation. In additon to the automation, such actions have been influenced by many other, especially economic factors, and the importance of individual factors is difficult to single out.

It appears from the above discussion that the overall effect of automation on need of labor has been much smaller than that of the mechanisation in forestry and agriculture, and at the same time difficult to point out quantitatively. Because both of these processes have been present simultaneously, the effects of automation have been overshadowed by those of mechanisation, while they additionally have been almost impossible to distinguish from the effects of other technological developments and general expansion of the industry.

2. Attitudes and Standpoints in Regard to Automation

Despite its radical social effects which certainly did not pass unnoticed, the mechanisation of the primary production never raised much of debate in Finland. The reasons of absence of such a discussion have not been comprehensively analyzed, but some conclusions are obvious. - One of them was the highly increased production per worker reached by mechanisation, while the work did not become heavier and often became lighter. Other reasons were the availability of jobs elsewhere, and generally higher salaries to both those who moved to cities or abroad and those who remained with the primary production and accepted the new methods of work. The trade unions did not oppose the change either, obviously because the income increased to their such members who remained in the trade.

Since the effects of automation were much smaller and difficult to distinguish from the general technological and industrial change, it is understandable that they were not much observed outside the individual plants and did not raise a general debate or discussion. The trade unions certainly monitored the progress of industrial automation but remained neutral, also in those branches in which the overall change was less radical, like in machine shops. From Mid-70's onwards they were prepared to accept a cautiously implemented automation, considering the increased efficiency of production beneficial to the workers and not only to the enterprises.

This standpoint was strongly correlated with the view of the general public on automation. Due to the absence of an active opposition and of substantial examples of adverse effects, the public took, at least from late 70's onwards, the automation as a natural step of the technologial development. Even emotional, adverse statements by individuals were very rare. It therefore appears that the attitudes towards automation are in Finland more neutral than in most other industrialized countries.

One factor which seems to have influenced the attitudes of the labor market parties in particular, are some academic studies which in early and Mid-70's were made on technolocigal automation and its effects in individual industries. In one such study, which dealt with workers and work environment in mineral industries, the effects were found generally favourable in dependence of level of automation, in about all respects examined. A decrease of number of workers was observed in one case only, due to introduction of on-line analyzers and to resulting elimination of shift work in laboratory (Niemi, 1975; Niemi et. al., 1977). Another study on work in metal industry indicated, depending on the type of production, the well-known U-shaped relationship which indicates a high job satisfaction for the pre-industrial manufacture by individual craftsmen, a low satisfaction for the manual work in transfer line type of work and a higher satisfaction again for more automatic production, but did not reveal major negative effects.

An ample attention was paid to the automation in the formulation of the national technology policy. Although some related actions had been taken earlier, this was really started through the appointment of the State Committee on Technology in 1979. This committee was set up (slightly condensing) "to analyze and evaluate the technological development and especially the automation; it was also to formulate proposals on decreasing the negative and supporting the favourable effects of automation- related technology, and an advencement of technology and of its use in production". The membership of the committee included representatives of all political and labor market parties, state administration, universities, research institutes and concerned interest groups.

After the commissioned, quite comprehensive analysis and evalution, the committee formulated a number of general and more detailed suggestions and recommendations on the economic policy, working life, education, research and development, transfer of technology, and information of technology and its effects. The resulting report

(1980/1982) was favourably received by the society. All publicly expressed opinions accepted the main recommendation of the committee which was a fast application of new technology. Practical actions were subsequently initiated by government offices and the governmental technologicy policy has followed similar lines since then.

The main deficiency of the report which was mentioned in some statements on it was the too limited discussion of the more extensive social and cultural effects and of their relation to alternative scenarios of future. However, such a criticism was not continued and was therefore soon forgotten, and no other body was charged later with a related task. It thus appears that both the national technology policy and the public opinion continue to accept the automation without criticism, as an inevitable step of technological development and economic growth which for their part are considered necessary. Parallelly to the growth of the acceptance, the ideas on the content of automation have become more diffuse, and many people understand it today as just another name of the technological development.

3.   Progress of Automation and its Interface with Science

Exploiting the expanding supply of technical means and wanting to have a tighter control of both details and totality of production, the industry invests continuously more to the most modern control gear. Thus the process plants are today provided with plant wide, digital, distributed automation and communication systems of their second generation. As such standard systems have been introduced, they have implied an almost exclusive use of a limited selection of basic control algorithms. Such standard systems do not favour the use of individual algorithms, but they seem to produce a control of sufficient level in practice. This development obviously does not support a transfer of results of control theory into industrial application, despite the fact that e.g. adaptive controllers have also been commercialized and program packages for computer aided design of control systems have reached the marketplace (Marttinen et. al., 1990).

Also in discrete goods production, the automation has resulted in large changes of the control of manufacture and of the nature of work, but generally it has been less penetrating than automation in the process industries. This is due to the dominating type of production; especially the manufacture of articles produced singly, like that of ships or paper machines is not easy to automate. Certainly the introduction of flexible and computer integrated manufacturing concepts has affected the efficiency and organization of work, the frequency of robots is higher than that in most OECD countries etc., but e.g. the challenges offered by aerospace industries to advanced control theory and technology have been largely missing.

Thus the progress of technology including automation has implied a fast and extensive acceptance of new standard tools, while the incorporation of PID-algorithms and other basic controllers of the standard software into process flow sheets and their tuning has been based on experience and experimentation. The increase of the available computing capacity has thus resulted more in the management of an increased number of functions from a central point in the plant through flexible user's interfaces than in an essentially higher theoretical level of control functions. In fact, even the classical methods of control theory have always been relatively little used for design of controls of industrial processes. The control related research has in Finland often concentrated more on the subject process and its modelling than on control with theoretically advanced algorithms, and has that way produced an appropriate contribution to new, industrial applications of automatic control and thus to a wider diffusion of automation. But this does not remove the fact that the gap from new achievements of control theory to practical implementations is persisting.

Several reasons can be stated to the fast progress of the more qualitative technology and to the existence of the gap. One of them is the scarcity, due to the relatively small size of the enterprises, of such industrial development units in Finland which are able to conceive results of modern science and convert them to solutions of practical problems. Another reason is the allocation of governmental support for technological development to universities in accordance with direct interests of industrial enterprises. Such rapidly increased money has resulted in a shift of interests of many university laboratories towards industrial problems of minor scientific importance, solutions of which may be useful to a narrow range of technology or even to a single enterprise only. This is one feature of the national technology policy of the 80's which affects the automatic control even more than other technologies which are better supported by industrial R & D bodies. Still one reason is the engineering education which tends to be more industrially oriented than that in most European countries and certainly more than that in the U.S. and Britain. While an evaluation of the automation related research in Finland found this generally quite satisfactory, the possibility of the student to make his/her Dipl. Eng. thesis work in the industry while being hired by the enterprise for that work, was found to affect unduly the research interests of the university (Rosenbrock et. al., 1986). Certainly this practice which presently applies to about 50% of all theses, has made many graduates to underestimate academic research and theoretical instruction.

Although purely scientific research is with evident success pursued by most of the regular staff in universities, it may in priciple be required to include multidisciplinarity due to the mere assumption of the feedback loop. If the controller is studied and developed scientificaly, also the process component should be approached, i.e. analyzed and modelled scientificaly. Because it belongs to a different technology, a cooperation of control theoreticians with specialists of that technology is called for. Problems of multidisciplinary research and education have been discussed separately (Niemi, 1988), and according to some experience the most helpful partners are usually found in another university laboratory representing the other technology. Such contacts may then lead to an extension of the co-operation to include also industrial partners, or a bridging of the stated gap which often has proven too wide to pass directly.

4.   *Discussion*

Basic automation has in Finland proceeded parallelly to mechanisation of the primary production, and its social effects have been effectively covered by those of the latter process. Its connections with the national structure of industry are evident which makes international comparisons difficult.

The publicly expressed, inofficial and official opinions on automation are favourable or at least neutral. Its advance is fast and takes benefit of modern computation and communication technology. Its contacts with background science are loose, which is partly due to the multidisciplinary nature of industrial control problems.

While only industrial automation has above been subjected to discussion, the computerization of office work, also called office automation, is in progress in business and administration. Due to the high number of employees in the service industries sector, it may have considerable social consequences.

BIBLIOGRAPHY

Marttinen, A., Niemi, A.J. (1990). CACE in Finland - A survey. In: Ravn, O., Niemann, H. (Eds) Proc. of Nordic CACE Symposium. Tech. Univ. of Denmark, Lyngby, p. 3.1.- 3.7.

Niemi, A.J. (1975). Effect of automation on operational technology and workers opinions in concentrators. J.S. Afr. Inst. Min.Met., Vol. 76, p. 26-33 (Special issue on recent advances in mineral dressing).

Niemi, A.J., Nuotio, E., Häkkinen, S. (1977). Level of automation and nature of work in concentrators. In: Lancaster, F. (Ed.). Proc. 2nd IFAC Symp. Automation in Mining, Min. and Metal Processing. S.Afr. Council for Aut. and Comp., Pretoria, p. 497-505.

Niemi, A.J. (1988). Introduction to the systems approach at university. Eur. J. of Engineering Educ., Vol. 13, No. 3, p. 263-269.

Rosenbrock, H., Balchen, J., van Cauwenberghe, A., Spang, A. (1986). Evaluation of basic research in automation technology in Finland. Academy of Finland, Helsinki, Publ. No. 2/86, 33 p.

-, (1980/1982). The report of the Finnish Technology Committee. Office of the Prime Minister, Helsinki, Publ. No. 1/82, 281 p.

Prospective Autonomy of Automation

J. L. Elohim
Mexican Association of Systems and Cybernetics
Cuauhtémoc, D. F., México

ABSTRACT

A cybernetic interpretation of essential "behavioral" possibilities of man; the natural being that has "acquired" or "developed" the "highest" transcendental autonomous functions; may lead the human mind towards the configuration of a prospective framework required for a precise determination of feasible autonomy's degrees in every particular domain of the technological world, which means to deal consciously with automation

Every tool, device, instrument, apparatus, machine,... i.e. every technological "entity" or "process" invented by men seems to arise basically from information acquired about some particular aspects of natural and/or artificial (technological) things or phenomena that they (the men) were able, either purposefully or eventually, to assimilate, usually while they were engaged in searching how to survive physically, affectionately and/or intellectually.

Recently, large intellectual endeavours aiming to get scientific knowledge of the real world, with the support of mathematics and physics, brought forward important advances that made possible the emergence of favourable circumstances in different fields of engineering for the "creation" of **automatic entities and processes**, i.e. self-moving, self-acting, self-regulating, self-governing,... artifacts which presumably could "behave as self-determining" subjects, although they were unable to realize that were restricted to perform exclusively in the particular environment assigned specifically for each of them. They couldn't be aware of their own self-determining possibilities.

Unavoidably every technological thing or phenomenon is a reflexion (like the image of an object in a mirror) of something that can be distinguished from other things; of a thing that is relatively autonomous in the earthly Nature increasingly altered by the dynamic presence of innumerable artifacts that are inserted here and there, from time to time.

It has happened similarly with the "automatic subjects" mentioned above. They wouldn't be what they are if some members of the scientific and technological community had not discovered the cybernetic principles in some peculiar aspects of the real world; if they had not begun to develop their philosophical nature, their epistemological characteristic, their teleological perspective and their methodological requirement, even though they have not usually worked very explicitly in this affair.

These are facts that have allowed the "successful" emergence of **automation** as such because it couldn't be ignored that many automatic artifacts and processes became parts of the real world already in ancient times and have reproduced successfully since then, even though their "performance" was not called automatic in those times.

So far, automation has been largely, although gradually, developed through the cybernetic understanding of possible effects of negative feedback connexions in networks and arrangements constituted by mechanic, pneumatic, hydraulic, electrical and electronic elements made from objects taken extensively from the so called inanimate world.

These automatic entities are capable to self-determine their performances exclusively in particular environments and situations, either chosen or organized or built on purpose. So far, an automatic artifact cannot perform successfully in any environment other than that or those previously specified when it was conceived and designed. Besides, to all appearances, such artifacts neither can evolve by themselves nor can search on their own whether they could develop a "higher" or a different autonomous possibility

However, a large number of technological processes, developed and implemented in the conventional domains of chemistry, biology, physiology and psychology, have provoked many unexpected effects in plants, animals (men comprised) and their environments; they are in fact side-effects that, could be said, have arisen "automatically" inside these living things.

Besides the study of the whole performance of every kind of living being by means of such approach, although it still remains in practice an open question, seems to offer better chances to get a good comprehension of autonomy and its prospective transcendence in Nature.

These two questions are potentially an immense panorama for the development of automation, if and only if they are tackled through a cybernetic approach.

To use such approach, for tackling a particular question, is to interpret cybernetically the dynamic features of the "entity" concerned in order to provide a whole understanding of its movement in the time and in the space, after assuming that such movement is the outcome of several feedback connexions (positive and negative) among the elements involved.

A very first requirement for the proper implementation of this approach when intending to deal with the development of automation is to configurate a framework that may provide an answer to a fundamental question: **What does it mean for a particular artifact to increase the degree of its autonomous operation ?**

A proposal for such framework is the cybernetic interpretation of man's potential performance which can be considered to be the latest outcome (not necessarily the last one) of the evolutionary history of the Universe -as men view it from the Earth- which so far makes possible to distinguish six major transcendental emergences:
 * Inanimate matter from ? by means of feedback regulators
 * Life from inanimate matter through adaptiveness i.e. *a posteriori* concern.
 * Prospectiveness in living things by means of *a priori* concern.
 * Sociability in the world of animals through indirect contact with the reality.
 * Awareness of oneself in some animals by means of cultural bonds.
 * Human consciousness or awareness of relative wholeness through ?.

When each of these events happened, although it is not yet known why...?, how...?, when...?, where...?, ...each one happened, it seems possible to assert that phenomena that never before had occurred and natural laws that did not previously exist, made their appearance and certainly some parts of the reality concerned increased their autonomy concerning the rest of the reality. Each transcendence being the "cause" of the appearance of more autonomous things, made "necessary" new properties to come into being:
 * Physical conformation.
 * Excitability.
 * Protopsyche.
 * Symbolic communication.
 * Concrete thinking.
 * Abstract thinking.

The whole panorama of Evolution's history seems to be reflected in every human being when the potentiality of his (her) performance is interpreted cybernetically (Fig. 1). May this view of **human autonomy** support the development of **automation** through the systematic search of possible **automatic artifacts that, on their own, develop further their autonomy.**

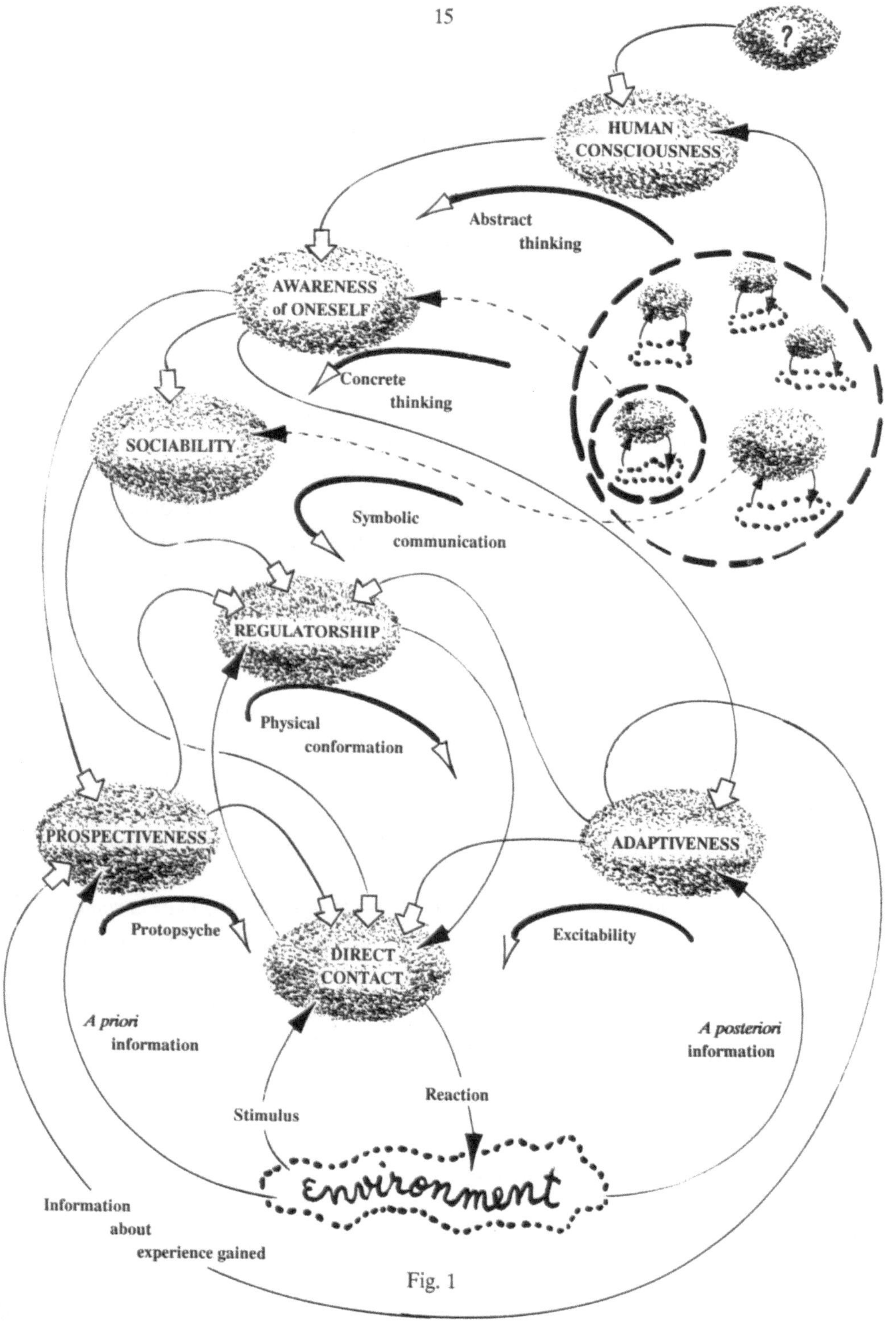

HUMAN CONSCIOUSNESS

?

Abstract thinking

AWARENESS of ONESELF

Concrete thinking

SOCIABILITY

Symbolic communication

REGULATORSHIP

Physical conformation

PROSPECTIVENESS

Protopsyche

A priori information

ADAPTIVENESS

Excitability

A posteriori information

DIRECT CONTACT

Stimulus

Reaction

environment

Information about experience gained

Fig. 1

A Rational Basis for Determining an Appropriate Level of Automation Technology

N. Learmont and M.G. Rodd
Department of Electrical Engineering
University of Wales
Swansea, UK

ABSTRACT

Whilst Automation Technology has been developed by Industrialized Countries, scant regard has been given to the impact of such technologies on the developing countries of the Second and Third Worlds. In some cases the direct, or indirect, effects of this automation have driven the local economy to the brink of collapse, as can currently be seen in many African states, as well as most Eastern European countries. Where once thriving, economies are now losing out to the advanced technology of the industrialized world. "...with transport as it is now, a shirt made in Spain by a fully automated process, can be delivered into the streets of Mbabane (Swaziland) before, and at a lower cost than, a locally hand-made one." (Rodd, 1991).

Since the goal of any manufacturing activity is to generate income either to the benefit of the individual or the state, industries in developing countries must be made competitive. The problems, though, of establishing a manufacturing base largely revolve around the selection and justification of production technology to be used.

In this paper a computer model has been developed to represent the manufacturing and distribution-network for a product, covering raw materials to retail item. This model, can be used as a ratonal basis to determine whether it is competetive (i.e. profitable) to manufacture a product in a given country and then transport it is competitive (i.e. profitable) to manufacture a product in a given and then transport it to where there is a real market demand.

1.   Introduction

The current state of industry in many developing countris has been likened to that of the now-advanced countries, but looked at two hundred years ago! (Stewart, 1978) Following on from this hypothesis, it seems unreasonable to suggest that a 3rd world industry could ever hope to be competitive with one from an industrialized nation. This is especially true when considering the impact that automation technology has had on the economies of many countries in both the Thrid World and Eastern Europe. This is, as highlighted by Svedberg (Svedberg, 1991), demonstated by the falling share of Sub Saharan Africa in world trade exports (from 3 % in the 1950´s to 1.2 % in 1987).

The aim of this paper is to postulate that such fall in trade figures can be at least be stabilised, if not actually reversed, by the application of an appropriate level of automation technology in a specific country being studied.

The concept used here is a simple engineering one: that one must find a basis for the selection of technology; not based on some sociological mis-understanding of what technology might be appropriate, but based firmly on simple engineering facts. These facts, in the end, all boil down to economics.

The basic assumptions of this approach are that there has to be: (i) an (international) market for the manufactured products; (ii) an appropriate distribution network; (iii) no prohibitive import/export duties. (This last point, unfortunately, effectively excludes some countries form importing manufactured goods from developing counstries). From these simple, but profound, assumptions, a simple computer model is suggested which will represent the manufacturing and distribution network for a product. By applying the appropriate parameters to this model and analysing the indications given, a clear, meaningful understanding of the problems facing possible technological solutions can be seen. Based on these indications, a rational engineering approach to the design of production processes can result.

2.   The Model

The model is essentially based upon the simple Japanese philosophy:

Profit = Selling Price - Manufacturing Price !

Whilst this is a totally simplified approach to economics, it embodies the fact-of-life that the only reason to manufacture something is to make a profit - be it profit to the indivudual, corporation or country. The computer-based model has been employed so that it can be applied to any manufacturing industry in any country, for trade with any other country at a single point in time (Fig. 1). The model determines wheter it is competitive (i.e. profitable), to manufacture a product in a given country, and includes consideration of transporting the article to where there is a market demant, etc.

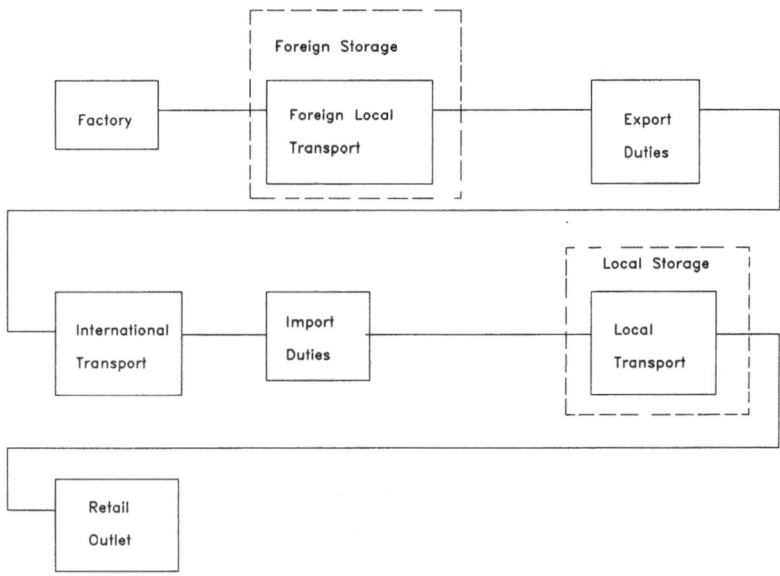

Fig.1. Schematic of the Manufacturing and Distribution Network.

The following basic variables are included in the model.

r = cost of raw materials / unit produced
l = cost of labour / unit produced
o = cost of overheads / unit produced
ft = cost of foreign local transport / ton
it = cost of international transport / ton
lt = cost of local transport / ton
e = export duties / unit
i = import duties / unit
fs = foreign storage cost / m²
ls = local storage cost / m³
us = unit size (m³)
uw = unit weight (kg)

From these the unit transport cost (T), the unit manufacturing price (M) ant the unit profit (P) are found for a fixed unit selling price (S) essentially by using the following basic relationships:

$$T = (ft + it + lt) * uw / 1000 + e + i + (fs + ls) * us$$
$$M = r + l + o$$
$$S = \text{fixed value}$$
$$P = S - (M + T)$$

3.   A Case Study

This example considers the position of the car radio manufacturing industry in the (theoretical) developing country of Swathawapali, and its attempt to compete with the U.K. for the car radio market in France. The figures used are representative only. If we assume that the Selling Price (S) of a car radio manufactured in Swansea (U.K.) and transported for sale to France, is S 60, this would also have to be the maximum Selling Price for an identical one manufactured in Swathawapali, for the same market. Obviously it is assumed here that the quality etc. of the two products is identical. The plant has a troughput of 80 units per day (the plant also produces a range of other products). The labour cost per hour is S 2 and hence the cost per unit was S 0.20. The raw material cost per unit is S 20 and the cost of overheads per unit S 7. Therefore the unit manufacturing price (M) ist S 27.70.

Now if we assume that the cost of foreign, local transport per ton is S 80, and the cost of international transport per ton is S 1000, and the cost of local transport per ton is S 100, we can analyse the cost cycle. We assume an export duty of 20 % and an import duty of 30 % are applied, to give the cost of export duties per unit of S 12 and the cost of import duties per unit of S 18. The foreign and local storage costs are taken as monthly costs of S 10 and S 15 per (metre)^3 respectively. The unit size and weight of the car radio, including packaging, are 0.002 m^3 and 0.75 kg. The unit transport cost (T) is thus:

$$T = (80 + 1000 + 100) * (0.75/1000) + 12 + 18 + (10 + 15) * 0.002$$
$$T = S\ 30.90$$

Therefore the unit profit will be,

$$P = 60 - (27.20 + 30.90) = S\ 1.9$$

This implies that the Swathawapali car radio industry can make a 3.2 % profit, if the selling price is S 60. However if the selling price fell below S 56.17, the manufacturing process would become unprofitable!

4.   Information from the Case Study

If a system under investigation is shown to be unprofitable. or a higher profit level is required, the model suggests where to look. This could be, say, by improving the distribution network or adjusting labour costs by, possibly, introducing certain levels of automation - which could range from simple machine tools to complex flexible manufacturing systems. To improve the distribution network, the key variables include unit size and unit weight. If these variables are able to be controlled, then suggestions can be made whether to send the unit by airmail, by airfreight, or by sea container. This again will depemd on volume - which goes back to unit parameters.

Turning to technological aspects, and particularly in the Electronics industry, the approach taken for rationalizing the level of automation technology can be based on the fact that in this industry, 50 to 75 % of the total manufacturing costs of a product lies in assembly (Lotter, 1984). Clearly, the cost of assembly essentially depends on labour (Elbracht and Schacher, 1984). Hence, simply, the higher the labour cost, the higher the level of automation technology required. The model has been designed to find the "appropriate" level, this being defined as the level of technology required to make the system competitive.

To do this, the model fist assumes that the entire assembly process is manual and if the system is un-economic, the computer will step through the levels of automation, until the unit profit becomes positive. This process hence determines the lowest form of automation technology required to make a profit. Of course, the model also takes into consideration capital depreciation, maintenance costs etc. as part of the overall cost of overheads per unit.

5. Conclusions

In summary than, this simple model can be a very useful aid to determining, on an economic basis, if it is possible for a country to produce a specific item. Th analyses involved provides useful clues as to the appropriate level of automation, and based on this, a rational basis for technology selection can be made.

6. Acknowledgements

The Authors wish to acknowledge the assistance of the University of Wales, Swansea in supporting the work discussed in this paper.

BIBLIOGRAPHY

Stewart, F. (1978). Technology and Underdevelopment. 2nd Edition. p. 60 - 61. Loncon. Macmillan.

Svedberg, P. (1991). The Export Performance of Sub Saharan Africa. Economic Development and Cultural Change. Vol. 39, No. 3, p. 550, April 1991

Rodd, M. G. (1991). Report To: The British Council on Engineering at the University of Swaziland. March 1991

Lotter, B. (1984). Automated Assembly in the Electrical Industry. Programmable Assembly. p. 75 - 92. IFS (Publications) Ltd. U.K.

Elbracht, D. , Schacher, H. (1984). Automatic or Manual Assembly? Boundaries of Economy at Middle or Low Batch Production. Programmable Assembly. p. 305 - 314. IFS (Publications) LTd. U.K.

The Impact of Information Technology on the Culture of Counselling

M. Molnar
Institute for Technological Design and Integrative Sciences
Technical University of Vienna
Department for Sociological Theories of Computer Sciences
Vienna, Austria

ABSTRACT

This paper descirbes the design of a computer-based adventure-game for use in vocational counselling of girls. It discusses the theoretical ideas and principles underlying the design and their translation into a functioning computer-program. It explores as well the cultural implications of using such a program for vocational counselling.

The paper draws upon the experiences of an ongoing project on ´Media-supported, activating vocational counselling' for young girls which has been commissioned by the Austrian Ministry of Labour and Social Affairs. The output will be an adventure-game on computer. The first three steps of this project have been completed: a) elaboration of a theoretical framework on base of empirical research, b) production of a script and its transformation into storyboards in coopertion with a graphic artist and c) development of a pilot version on Macintosh. The implementation of the program in real user-situations and an evaluation are planned.

1.   Idea and theoretical framework

Counselling-practices are traditionally strongly influenced by school-based models of learning and knowledge transfer. In contrast to this new counselling-practices offer a more informal atmosphere, e.g. providing free access to open-house-centres, where boys and girls can get information related to working-life from video-clips, brochures and computers. They can sit together in peer-groups or communicate directly with one of the vocational counsellors, if they wish to. In addition to a cognitive approach relying on information and facts a more playful and active access to dealing with working-life should be made possible.

In that context the idea of an interactive adventure-game on computers can be placed. In the following some basic didactic principles underlying the design of the game are described.

1.1. Didactic aims

Ideas of boys and girls about occupations are often misty, idealized and based on little factual information (Egloff, 1984). Young people do not have the experience and tools to look at all different aspects of vocations or to ask the right questions in counselling-situations. The first aim is to help them to develop a more concrete view of occupational life and to give them a guideline for their own thinking (BMAS, 1989).

The second important principle is to expose young people to criteria along which they can organize their own decision-making process. In "learning by playing" they should develop a set of differentiated criteria and apply these to their own decision-making. The focus is on supplying process know-how and not on knowledge about certain facts (Egloff, 1984).

1.2. Target group

The game especially focuses on girls. This is because girls´ entrance into working-life and a career is often handicapped and problematic (Lechner et al., 1991; Toth, 1990). Program structure and user interface are based on software-ergonomic principles (Shneidermann, 1987; Fähnrich, 1987). Beginners with no experience should be able to find their way through the game without frustrations. The pedagogical style of the design (topics, language, optical and acoustical presentation) establises an intentional connection to forms of media-reception commonly found in youth cultures, for example in comics, animation-films and video or computer-games. Pictograms guarantee a direct relation between seeing and immediate understanding (Hülsmann, 1985).

Playing with the computer should not be a single, isolated action. This is why a general media-concept for the best use of the program was developed. The intention is to support the transfer of know-how about the process of decision-making from the vocational situation into real life. Reflection on and handling of that know-how should be supported during and after the game. Additional action might include user-sessions of groups, offering outprints after the game, initiating communication about the game with teachers, parents, vocational counsellors and, furthermore, to give boys and girls instructions of how to transfer their knowledge to their own individual sitation. As found out in psychological media-research (Burkart, 1983), the didactical influence of the medium can be strenghtened by a follow-up communication.

2. The adventure-game "JOB-HOP"

JOB-HOP is built around three girl-protagonists: the secretary Karin, the hairdresser Tina and the electrical mechanic Eva. They are identification figures, that lead the users through the program. They invite boys and girls to follow them to their places of work and as them about a variety of aspects of their employment. The picture shows them in front of the entrances to their places of work (Fig.1.).

Fig.1. The three girl-protagonists Tina, Eva, Karin

The occupations were selected according to these reasons: to make the software attractive for girls it was necessary to offer well-known and often chosen vocations. On the other hand it was also necessary to present less known job-alternatives. The important point is, that all three examples illustrate more general structural aspects of employment and career. It is not the objective of the game to lead to a decision pro or contra one of those special vocations.

Fig.2. Eva´s place of work and the screen-mask

Information pertaining to each of these vocations is structured around several aspects, as typical work tasks and tools, income, occupational risks, career opportunities - all together seven points. Those aspects are the same for every example. Users can access information by click on one of the info-field-icons, demonstrated in Fig.2.

On the right side are the icons for each of the above mentioned topics of information. At the top three doors can be seen, symbolizing the entrances to each place of work. Users may proceed in horizontal or vertical steps. They may inform themselves about various aspects of one example or make cross-comparisons. In addition the special aspects of each vocation are captured so that users can move within 21 different situations.

Each of these situations contains two elements: a) presentation of information and b) interaction with the user. In the info-field "income" of the hairdresser for example. the user is asked, what a hairdresser might earn after five years in the job. Three pigs of different "volumes" appear on the screen, symbolizing a multiple choice-question (Fig.3). The user has to think over the question, has to decide on an answer and to compare it with the feed-back she/he will get then from the program.

Fig.3. Storyboard for info-field income/hairdresser

The game is no test, that would require correct right- or wrong-answers. Many of the questions are more funny than serious. That is one of the motivational factors to play the game. Boys and girls will be exposed to a variety of aspects with which to analyze working-life. They will reflect on them and discuss and laugh together. As part of a general didactical concept of vocational counselling this game cannot be a substitute to, but rather an enrichment of face-to-face-counselling.

3.   Changing the culture of vocational guidance

What´s new, alternative or different in integrating the adventure-game JOB-HOP into the practice of vocational counselling? Automation systems have been already used in the counselling services, but rather as fact providers. In contrast, JOB-HOP challenges traditional ways of using computers in vocational counselling. This will have an impact on social practices in particular

- on the product itself:

* The aim of the program is not the reproduction of single facts or information units, but the transfer of an implicit and abstract didactical idea to the user.
* A pedagogical design is communicated through animated cartoons, comics and conventional video-games.
* The user´s relation to the program is more affective than cognitive.

- on the user-group:

* The focus of interaction and reception stimulated by the program is rooted within the youth-culture itself.
* The game is used individually and self controlled. Users only decide on starting and ending the program, lenght of time at the computer and selection of information. There is no control by authorities.
* Using the game in groups will most likely reinforce its didactical value.

- on the counselling-situation:

Important points of reference for the use of JOB-HOP as part of a counselling process are:

* reinforcing the main message of JOB-HOP in face-to-face-communication,
* transfer of its contents into school-based learning about working-life,
* encouraging autonomous exploration of occupations in real life (for this purpose users are offered an outprint with questions and suggestion similar to the game).

These hypotheses serve as a basis for the empirical evaluation, which is planned together with the implementation of the computer-program in counselling-practice.

BIBLIOGRAPHY:

Bundesminist.f.Arbeit u. Soziales (1989). Tips zur Berufswahl. Wien.

Burkar, R. (1983). Kommunikationswissenschaft. Böhlau-Verlag, Wien-Köln.

Egloff, E. (1984). Berufswahlvorbereitung. Grunlagen, Didaktik, Unterrichtseinheiten. Lehrmittelverlag des Kantons Zürich, Zürich.

Egloff, E. (1984). Mein Berufswahltagebuch. Schülerteil des Lehrmittels "Berufswahlvorbereitung". Lehrmittelverlag des Kantons Zürich, Zürich.

Fähnrich, K.P. (1987). Software-Ergonomie. Oldenbourg-Verlag, München.

Hülsmann, H. (1985). Die Maske. Essays zur technologischen Formierung der Gesellschaft. Ver. Westfälisches Dampfboot, Münster.

Lechner, F. et al. (1991). Vergessene Frauenarbeitsbereiche. Focus-Verlag, Gießen.

Shneidermann, B. (1987). Designing the User-Interfaxe. Eddison-Weseley Publ., USA.

Toth, St. (1990). Berufsausbildung und erster Berufsstart. In: Friebel H. (Ed.). Berufsstart und Familiengründung - Ende der Jugend? Westdeutscher-Verlag, Opladen.

Historical and Present Situation of Automation in Czechoslovakia from the Cultural and Sociological Points of View

P. Vavrin
Department of Automatic Control and Measuring Technique
Faculty of Electrical Engineering
Technical University of Brno
Brno, CSFR

ABSTRACT

This article gives a brief overview of the main differences between cultural and social consequences of automation in Czechoslovakia and comparable Western European Countries.

In various theoretical works published a number of positive as well as negative phenomena are mentioned by which automation affects our life. To avoid misunderstanding let us note that attention will be focussed especially on automation of technological - mostly industrial processes. For the cultural and sociological areas the following features of automation are important:

1. *Positive effects:*

1. Automation, which is often referred to as the second scientific and technical revolution, liberates man from monotonous and tiring mental work (the first scientific and technical revolution concerned physical work).
2. Automation distances man from direct participation in the production process and thus people can work in a better working environment.
3. By optimizing the production processes and by strictly observing the required parameters, savings in energy and material are achieved, which again leads to an overall improvement of the environment.
4. Automation reduces production costs and many products become available at a reasonable price for the broad masses.
5. Automation reduces the proportion of unskilled or less skilled manpower in favour of more educated workers.
6. The part of population which is necessary to provide the necessaries of life is decreasing so that a greater part of society can devote themselves to the so-called tertiary sphere, which includes culture.

7. The application of computer technology requires a strict observance of a given order and regulations, which is of doubtless educational effect.

2. *Negative effects:*

A. By distancing from direct production traditional craftsmanships disappear. As a result of division of work the specialization of workers is narrowed down and so is their scope of interest and knowledge, which results in reduced general education level.

B. The absolute volume of production is increasing, which means a greater exploitation of natural resources and a greater threat to the environment.

C. Automation makes it possible to construct large power engineering and industrial complexes (power plants, chemical reactors, etc.), which always brings severe interference with nature and peoples life, with the risk of disturbing natural balance and giving rise to conflicting or even catastrophic situations.

In the above list of positive and negative effects the positive ones prevail at first sight. If, however, we assign certain weight coefficients to the individual effects, the result is no longer that unequivocal. The magnitude of these coefficients may be a subject for discussion and so we shall not dwell on this aspect. We shall rather concentrate on how that above theoretical effects found practical realization in Czechoslovakia which is a country of long industrial tradition. It must be borne in mind, however, that the process of automation started on a larger scale in the late fifties, when all spheres of life in Czechoslovakia especially technology and industry - felt the effects of totalitarian rule.

It can generally be said that any major projects of automation in Czechoslovakia (the same as in any other communist countries) were undertaken not to meet the needs, the customer's interest or technical urgency but to comply with the wishes of some member of the party or government hierarchy, or on the basis of a decision taken by a party or government committee.

Professional qualification or competency to take such a decision was relegated to the second place. Quoted as examples may be the construction of the nuclear power plant in Temelín or the hydroelectric project Gabfíkovo. These two constructions relate directly to point C above. For various technological reasons the construction of huge industrial or other complexes is of some advantage but if ecological and sociological counterchecks are suppressed, automation may indirectly become a highly negative and, in extreme cases, dangerous tool. In conditions of democratic rule decision taking and market economy, all the acting forces are in balance. To use the terminology of control engineering, we might say that the respective feedbacks are sufficiently quick and effective. Under totalitarian rule we often witnessed auto-oscillations. The causes of these events could be proved exactly by methods of the theory of dynamic systems.

Another deformation can be shown on the example of power economy. Instead of applying automation to the consumption of power, attention was focussed on its generation, which temporarily solved the problem of lack of power but the price was an enormous devastation of the environment (northern Bohemia).

Another frequent occurrence was the poor quality and imperfect final execution of automation projects (with the exception of a few systems imported together with foreign technology). With a view to safety and required operational reliability these systems were in most cases backed up by an equivalent solution so that almost all positive effects manifested themselves on a reduced scale or not at all. With regard to points 1 and 2 it should be added that workmen frequently and intentionally damaged various automation systems since these systems represented for them stricter inspection of their performance and sometimes even the loss of bonuses for working in harmful and hazardous environment. A weird example of this type of deformation can be seen in the case of operators of NC machines: workmen hired graduates to rewrite the programs in such a way that higher performance was achieved at the expense of the operational and safety parameters of the machines (higher feed rates, deeper cutting tool bites, etc.)

One of the promising projects for the future is the envisaged automation of telephone and telecommunication networks in Czechoslovakia. This project should make it possible to extend working activities at home, to provide access to information, courses and learning opportunities of all kind. Only then could the positive effects of points 5 and 7 be felt.

3. Conclusion:

With respect to the above deformations it must be said that the effect of automation on the cultural and sociological aspects of life in Czechoslovakia proved to be more negative than could be anticipated on the basis of generally known and theoretically founded rules. We hope that the changes in world politics, economy and commerce will also lead to the required changes in this area. From the position of our Technical University we are doing our utmost to help achieve this goal.

"East-West" Approaches of Automation:
a Case-Study

L.K. Blach, K.Cybinska-Pirog, J.B. Lewoc and M. Rozent
International Consulting and Marketing Co.Ltd.
Household Powered Appliance Works (Polar)
Leader (Leading Designer)
Institute for Power System Automation (IASE)
Wroclaw, Poland

1.   Introduction

On the IFAC workshop in Budapest, there presented two general designs of similar system: Badel - a data base for a coal fired power plant (Lewoc et al., 1989) and PRIMO/S a data base for a nuclear power plant (Stegbaurer, 1989). Both systems are big enough to produce symptoms characteristic to complex automation systems of major cultutral aspects. We considered it worthwile to analyse the differences in hardware and software of the systems and try to find some objective reasons for the differences.

2.   Comparison of Hardware and Software Structures

The hardware structure (Fig.1) of Badel is shown for the whole power plant though the subsystem for one power generation unit (PGU) is directly comparable to PRIMO/S.

When comparing the hardware structure of PRIMO/S (Fig.2) and Badel, the two major differences can be found:
-    in Badel, the hardware is of one rank less complex than in PRIMO/S: 8-bit microcomputers in DAQ-s and LSI-ll-s in the monitors comparing with PDP-ll control computers and Vax on the high level,
-    Badel uses the star configuration with V.24 links while a local-area network is used in PRIMO/S.

The basic difference in the software structure reflects the adopted approaches to the design work. For the polish system, the structure implied by the task and hardware environment. For the other approach, the real time kernel and a set of services are defined; these are used to solve the problems of various applications.

The first difference is implied by the economic condition of our country which leads to underinvestment in the computing science. The other differences are consequences of the first one: to achieve similar results on less complex hardware, the Polish designer has has to work more hand. And the results are as follows:

The star configuration is more robust than the one based on a local-area network (Izworski and Lewoc, 1991 a and b).

31

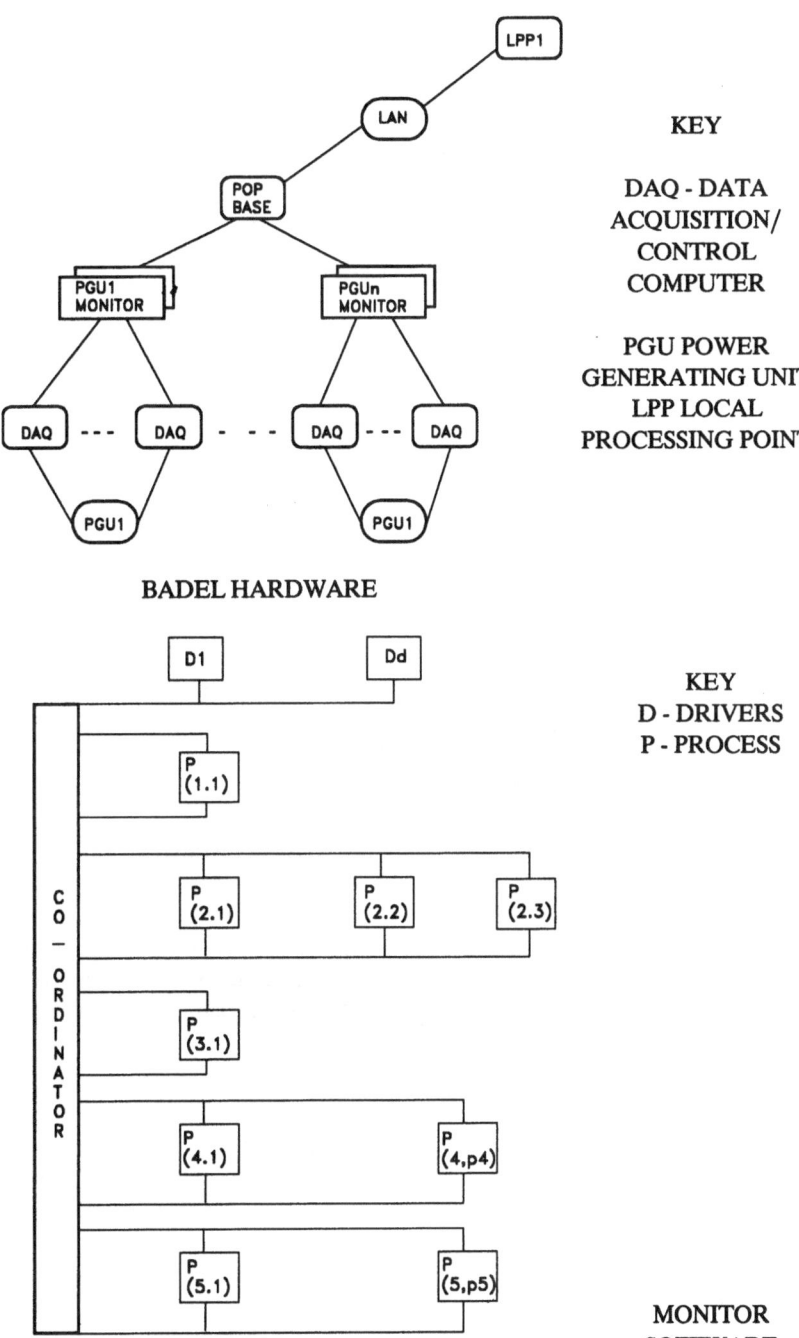

KEY

DAQ - DATA
ACQUISITION/
CONTROL
COMPUTER

PGU POWER
GENERATING UNIT
LPP LOCAL
PROCESSING POINT

BADEL HARDWARE

KEY
D - DRIVERS
P - PROCESS

MONITOR
SOFTWARE

Fig. 1. Hardware and software structures of Badel

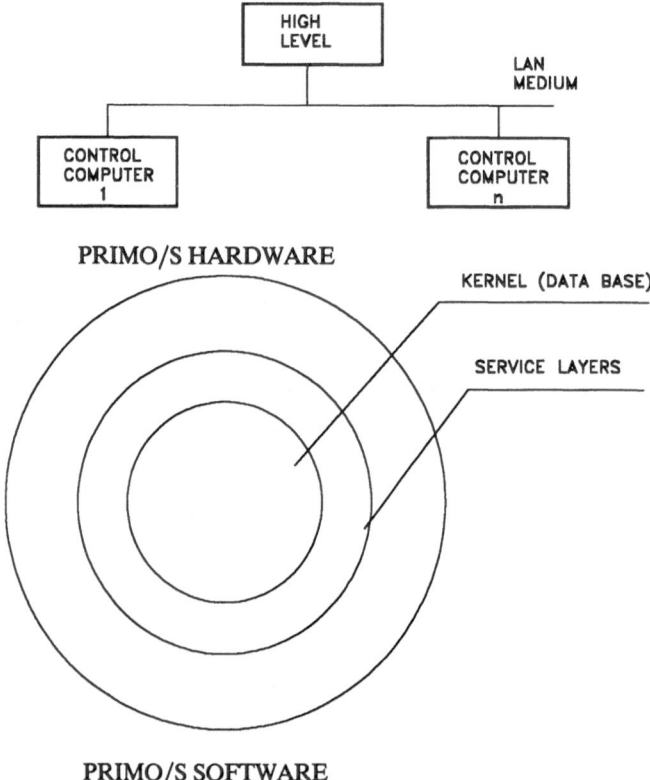

Fig. 2. Hardware and Software Structures of PRIMO/S

The layered structure of software in PRIMO/S requires general - purpose standards and interfaces between the layers which imply overheads for unnecessary exchange of data, interlocks, protections etc. The overheads in Badel are significantly lower and the latter software is much more effective.

3. *Some Cultural Aspects*

The first cultural aspects is somewhat heretical: in our opinion, Polish designers are better qualified than those of the West. We think here of the designers who achieved the professional success. It is because they have to work much more hard to get the success.

Another major cultural aspect is that the designer in Poland has to adopt in a maximum possible extent his technical means tospecific plants. Therefore, he must learn the plant in a much higher degree than his western competitors. In Polish conditions, the success of implementation simply needs the approach; I have to do my best and achieve as much as my means enable. This is often connected with limiting of personal demands to the level which can be met and not that implied by knowledge, experience, etc.

4. *Some Conclusions*

An initial analysis shows that the Household Powered Appliance Works "Polar" need a computer integrated manufacturing system providing facilities not available presently.

The above considerations lead to the conclusion that the system should be designed by Polish designers. A simple prediction of costs and benefits gives some economic evidence for it.

BIBLIOGRAPHY

Izworski, A., Lewoc, J.B. (1991a). Robustness Comparison for Distributed Control Systems of Various Topologies. In: DMCS, IFAC, Zurich.

Izworski, A., Lewoc, J.B., (1991b). Quality Comparison for Big Power Plant Monitors: A Case Study. In: RCSDu H^{oo} and RM, IMC/SERC/IEE/IEEE, Cambridge.

Lewoc, J.B., Slusarska, E., Tomczyk, A., (1989). A Distributed Database for Real-Time Control/Monitoring of Power Plants. In: DDRIC, IFAC/IFIP Budapest.

Stegbauer, W., (1989). Intertask-Communication Inside a Real-Time Database. In: DDRTC, IFAC/IFIP, Budapest.

Automation Success Factors as Seen from a Developing Country

J. Cernetic and S. Strmcnik
Department of Computer Automation and Control
Jozef Stefan Institute
Ljubljana, Slovenia

ABSTRACT

The most common problems and the corresponding success factors occurring in computer-based automation projects are discussed, as they are seen from within a department participating itself in research, development and applications in Slovenia. As a result of experience, recommendations and actions are proposed about where to get help and how to take remedial actions in automation problems.

1. Automation problems in a developing country

Let us begin with the problems in computer-based automation. From these, we will look in the next chapter for their reasons and where we can get help. In the third chapter, we are trying to sieve out the critical success factors. At the end, we give some general recommendations, as well as a proposal of specific actions, currently dealt with in Slovenia.

Our country has not been seriously exposed to computer-based automation until the early seventies. So we will present here a short review of our problems according to three seven-year periods, running up to 1991. In the first period, some rare computer automation projects were performed completely by foreign companies. A handful of domestic people, mostly enthusiasts, have been searching for proper knowledge in this emerging field, using some small support funds of the former Research Council of Slovenia. The critical problems in this period were: lack of technical know-how, inadequate education of potential users, very naive expectations and absence of true motivation for considering computers as potentially useful tools in automation. In this situation it was not surprising that the domestic users had a blind confidence in foreign bidders of automation equipment. In a couple of cases some of the users could state that this is not a good position; unfortunately too late, because the automation project has failed and the invested money has gone. Nevertheless, such sad experiences gave many impulses for the development of domestic professionals in this field.

In the following second period, respectively, some groups emerged in Slovenia, able to tackle simpler automation tasks by themselves. They developed and installed their own microcomputer-based systems, performing data acquisition, supervisory control and even closing some PID control loops. Along with this came the first domestic successes and, of course, the failures. In this time, there were still enough technical problems, but now also the nontechnical ones began to show their acuteness. For example, a project was recognized as failed after some years of undetermined "unfinished" state because of missing economic justification, although it was managed technically correct, the control system has passed the initial production tests and considerable effort was invested in its documentation. Some other projects were in fact never finished due to sloppy initial system specification and never ending "modifications" or "improvements". Lessons learned in this period include statements like the following ones:

- do not begin an automation project unless a thorough feasibility and techno-economical justification study has been done;
- prepare system specifications before you go on computer programming;
- do the documentation in parallel with system development, not afterwards;
- do not reinvent the wheel, use modules;
- apply some methodology for system development.

The third period of control system development in Slovenia is marked by considerable maturity of our professional community: the makers as well as the users. Both, in general, have acquired enough experience in previous years to be able to appreciate correctly the need for their own know-how. Also the expectations became much more realistic, partly due to great advances in computer technology and widely available (personal) computers. From now on more extensive and more complex applications of computer control came under consideration, some even including advanced control techniques. In this new situation, a new and deeper layer of automation problems came to the surface.

It was clearly recognized by potential users that computer control has an enormous potential in making the industrial production more profitable in terms of better product quality, higher production rates or lower energy consumption, as well as more acceptable for the environment and the human labor force. Unfortunately, we are confronted daily with serious obstacles, which render the introduction of modern control systems in our factories very difficult, in some cases even impossible. For example, production engineers having daily problems with obsolete technological equipment, trivial deadlocks in operation and poor working motivation of their staff, are - as a rule - not very enthusiastic with computer automation plans. The acceptance is much better in new or reconstructed industrial facilities. Unfortunately, in this case the investment resources are usually cut too low to afford a "normal" control computer system. It is sometimes said to the project team that "the computer will be fitted later ...". Often it is not, because market changes do not allow any upgrades.

The list of problems is even longer if the decision is reached to start the project. A particularly harmful group of automation problems which we experienced from this list, is related to management. It is clear that any project managed poorly will have little success, but an automation project lacking support from the higher management line has indeed very low chances to succeed. Very often, our managers expect that a modern control system will solve all production problems, or similarly that the project team alone has the responsibility for solving them. Another actual problem are partial interests of different kinds. In our country very often the users are divided in those who believe in domestic know-how, and those who would like to buy the complete control system abroad. It turns out that this, in fact, is the consequence of missing long-term national strategy. As long as this is not agreed on the highest (national) political level, domestic makers of control systems are inferior in negotiations when compared with foreign bidders, usually strong companies with worldwide reputation.

There is another group of implementation problems which demanded a great deal of effort for solving in some of our projects. First, there are problems in upgrading the measuring and control instrumentation, to which the process control computer should be connected. If there exists already some instrumentation in the process to be automated, it is relatively old and of different brands, so the signals sometimes have to be adapted to standard levels (mostly the 4-20 mA current loop). Further, the procedures for acquiring additional (new) sensors and actuators (which usually have to be imported) are somewhat complicated and slow. Often the project was blocked due to missing or wrong instrumentation. There are also problems with installation, because most our factories have not enough competent professionals to do it. Similar problems occur when other conditions for convenient and safe operation of the control system should be assured, as e.g. trouble-free process signal connections, proper electrical grounding, fail-safe power supply, provisions against disturbances, air-conditioning, etc. Due to such problems and circumstances, it is sometimes very hard to achieve a continuous and reliable operation of computer-based control systems. The problem of reliability is still harder if the process operates batch-wise and good control relies on data from previous (troublefree) operation, e.g. in batch pulp cooking.

2. *Where to get help?*

Bad experiences from the above-mentioned problems and the resulting pressures in application projects have stimulated an intensive search for available solutions. In recent years we could do this a little more systematically, so the present discussion of automation success factors is the result of these efforts. The statements mentioned here come partly from the careful study of selected references, and partly also from personal real-life experience in developing computer-based control systems.

It turned out that the intrinsic success factors, which are valid also in computer-based control, are extensively dealt with in the modern disciplines of systems and software engineering (Blanchard, 1991; Pressman, 1987). Therefore we have set out to gradually include this knowledge into our practical know-how, next to control engineering. In recent years we have seen from our own experience that the following areas can contribute almost immediately to greater success in automation projects: proper technical and economical justification of projects (Cernetic and Divjak, 1979), techniques of effective project management, and computer-aided software/systems engineering (CASE) methods and tools (Cernetic, 1990). We now use the respective techniques, methods and tools regularly in our projects.

What is more important, we know that, according to Livingston (Livingston, 1988), the human inability to deal with complex matters and situations is the most important single source of problems in modern automation projects. So in the future, we plan to pay more attention in control system development to human factors, as well as to strategies for solving complex problems.

Considering all the above, we are now able more correctly to define what makes an automation project successful. If a project should count as a total success, it must not only be finished within the planned schedule and resources, with the control system operating without remarkable bugs! In fact, the control system must satisfy its users in the long run, helping in solving the right production problem and proving its economic justification in its entire lifecycle.

3. A proposal of recommendations and actions

Deeper reflection on our automation problems and those of developed countries has shown that they are not entirely different. So we concluded that in the search for solution strategies we can rely heavily on the know-how developed abroad. Yet, in addition, we must consider our local circumstances and develop our specific recommendations and remedial actions. Here we wish to report about three such results. First, we will give a proposal of automation success factors formed from an informal inquiry in which about two dozen of professionals, involved in automation, either as users or makers, have been asked for their personal opinion about this topic. Second, the spectrum of activities of our department will be described briefly. It was formed some years ago in an attempt to suit local needs in Slovenia for computer-based control on the research, development as well as in the application level. Third, we will mention a proposal for organizing national resources in Slovenia, involved in process automation. Its intent is to get a wide framework for more efficient transfer of knowledge and positive effects in applying control systems technology on a national scale.

Automation success factors as seen in Slovenia.

After a process of necessary stylistic unification of answers from the inquiry, the automation success factors were arranged in two main groups, each having three subgroups. The structure is the following:

STRATEGIC FACTORS: OPERATIONAL FACTORS:
vision   organizational
education   technology
justification   human factors

The complete list of success factors gathered in our inquiry is given in the Appendix.

How we do it.

The Department of computer automation and control was formed five years ago from a group of control, electrical and chemical engineers, which have been working together previously on the same research and development institute. The department is active in research, development and applications in computer-based control of systems and processes, offering also corresponding engineering, education, consulting and other services. Our principal objectives are to cultivate and expand knowledge, to develop tools and building blocks which enable the use of advanced control in industry and other areas of life. The mission of the department is to help our country in bridging the proverbial gap between control theory and practice. It was designed to achieve this by an active interplay of research, development and applications within four specific areas:

> knowledge about systems control,
> procedures and tools for control systems design and implementation,
> building blocks for control systems, and finally,
> applied solutions of control problems.

In the first area, we are involved in research, development and application of principles, procedures and methods of system theory, control systems science and automatic control. We are particularly interested in system identification, optimal control and adaptive control. In the second area, we try to acquire the most actual knowledge and, at the same time, to develop and to use tools (mostly computer-aided) which successfully support the entire life-cycle of control systems development. We search for complete and practical solutions of control problems, using approaches, methods and tools of computer-aided control engineering. For this end, we currently follow the progress in computer-aided control system design (CACSD) and computer-aided software/systems engineering (CASE). In the third area of our activities, we develop special hardware and software as building blocks which support the use of advanced control methods. For example, a microprocessor-based multiloop controller, built in small series according to highest industry standards, was designed for cost-effective solution of medium-size advanced control problems, such as e.g. modern combustion control in steam boilers, industrial furnaces, etc. In the fourth area, we use all the above knowledge, engineering experience, tools and building blocks in solving complex computer-based control projects, as well as in solving particular and demanding control tasks. Mostly we are dealing with continuous and batch processes in the chemical and other processing industries.

A proposal for a national research and development policy.

From practical experience in the past few years, we can state that the above mission and the activity structure of our department were successful and are relatively well suited to the needs in our environment. At the same time, we see that this is not enough to arrive at significantly greater efficiency in advanced automation on a national scale. Therefore we started recently an initiative, together with our professional colleagues from university and industry, to form a national success strategy in the area of control systems technology. The basic idea of this initiative is, first, to convince our government that the intensive introduction of modern control systems technology into all segments of national economy must become one of priority goals.

The arguments for this proposal are mostly known to control and system engineers: i.e. the multiplicative positive effects which such measures can have on more efficient performance of different systems, no matter whether they are technical, economical, social, or whatever. In addition, such measures are usually very cost-effective and bring short-term, as well as long term results in improving the infrastructure of national economics. The second, i.e. practical part of our initiative, is how to plan and organize the activities, which would tie together and motivate proper subjects. In this sense, initial proposals, agreed between the most interested institutions, were sent to the Slovene government and its Ministry of science and technology.

In conclusion of this study about automation success factors we think that our viewpoint on this matter may be interesting and helpful for our professional colleagues, working in similar conditions. We know that engineers dealing with the development of computer-based control systems, to be really successful, must seriously consider not only the technical success factors, but also the nontechnical ones. In this sense, we believe that the mission of our department and a sound national strategy can help our country significantly in mastering the technology of control systems (which includes both, control science, as well as control engineering). This, in turn, will lead not only to greater success in automation, but also more generally to better development of our national economy which right now is the most critical factor for national survival in Slovenia.

BIBLIOGRAPHY

Blanchard, B.S. (1991). System engineering management. Wiley - Interscience, New York.

Cernetic, J. (1990). Suitability of CASE methods and tools for computer control systems. Informatica (Ljubljana), Vol. 14, No. 1, p. 38-44.

Cernetic, J., Divjak, S. (1979). Justification of process computer control. Vestn. Slov. Kem. Drus., 26 (3) p. 325-342 (in slovene).

Livingston, W.L. (1989). A process of elimination. Control Engineering, April 1988, p. 154-160.

Pressman, R.S. (1987). Software engineering- A practioner's approach. McGraw-Hill, New York, p. 32-79.

APPENDIX:

Automation success factors compiled from an informal inquiry in a group of professionals in Slovenia, 1991

I. STRATEGIC FACTORS

A. Vision

1. Conscious need of technology development
2. Belief in high profitability of automation
3. Motivation by market competition
4. Business strategy related to (supported by) automation
5. Realistic expectations concerning possibilities and effects
6. Long-term and flexible automation plans
7. Integral approach and correct goals
8. Vision supported by a realistic implementation plan (strategy of small steps)
9. Impartial management support
10. Consideration of evolutionary nature of computer control projects
11. Previous feasibility study and project justification

B. Education

1. The user must know the possibilities of computer control technology
2. Readyness to invest in education
3. Education at all levels of organization
4. Higher technical and organizational culture
5. Cultivation of interdisciplinary knowledge
6. Knowledge of organizational skills
7. Systematic accumulation and evaluation of experience
8. Step-wise execution of a project gives the users better chances to follow
9. Knowing the basic engineering principles and the dynamics of the process (system) to be automated

C. Justification

1. Clear business situation of the investor (user)
2. Stable (or at least foreseeable) market situation
3. Defined criteria for objective estimation of automation benefits and costs
4. Consideration of different bidders
5. The economic efficiency of the project has priority over its technical excellence
6. Fast project execution gives shorter pay-off period
7. Funds for production testing must be assured
8. Economic effects of automation depend also on some uncontrollable factors and therefore can be achieved gradually

II. OPERATIONAL FACTORS

A. Organizational

1. Flexible organizational structure
2. Involvement of an important (high-level) manager
3. Good project planing and management
4. Interdisciplinary project team
5. Tight cooperation between system developers and users
6. User organization assigns a special group to be responsible for automation
7. Good coordination between different groups participating in control system development
8. Informal organization within a particular working group
9. Choice of a reliable equipment supplier
10. User staff fully participating in the project must be free of routine production tasks
11. Proper organization and procedures for solving current problems in development
12. Established methods, measures and procedures for the evaluation of project results
13. System installation, testing and maintenance in close cooperation with the group responsible for measuring and control instrumentation
14. The user takes care of system maintenance
15. Steady monitoring, improvement and evaluation of effects achieved with automation

B. Technology

1. Use of new but proven technology
2. Integral approach to system design
3. Step-wise transition to brand-new advances
4. Inclusion of control considerations into process flowsheet design as early as possible
5. Use of proven modular solutions
6. Organizational changes before the introduction of new control technology
7. Avoid automation of a technologically outdated process
8. Tidy documentation of the process to be automated

C. Human factors

1. Clear motivation of all participating partners
2. Distinct lines of responsibilities
3. Management ready for organizational changes
4. Project staff adhering to the goals of his organization
5. Different working style is usually needed with automated systems
6. The control system has to be simple for use
7. Specific working conditions in developing countries must generally be considered in automation planning

Anthropocentric (or Human-centred) Technologies in Europe
- Experiences and Perspectives -

D. Brandt

Department of Cybernetics and Engineering Education, (HDZ/KDI),
University of Technology (RWTH)
Aachen, Germany

ABSTRACT

In this report, three case studies are presented which describe anthropocentric systems implemented in different European countries. These cases are discussed with regard to the term "anthropocentric". Furthermore they are discussed with regard to their economic viability as well as with particular reference to the specific conditions of Southern and Eastern Europe.

1. The term "Anthropocentric Technologies"

Recent developments of technology seem to have shown a strong orientation towards finding the best technical solution to a problem. But frequently they show little consideration of people concerned. This approach may be called "technocentric". In contrast to this approach, we suggest to look at problems with the aim of finding a solution which balances advantages of a merely technical solution, and the needs and competencies of people. Several terms have been suggested to describe this approach: Human-centred, or skill-based - or "anthropocentric". This approach includes to look for economically viable solutions. Thus, Anthropocentric Systems concern people, organisation, technology in their societal and environmental context. According to these issues, five questions may be asked to evaluate technology in terms of its anthropocentric features:

i.   Has technology been designed taking into account people at their individual workplaces? (This question comprises issues of health, safety, ergonomics; furthermore skills, needs and motivations of people)

ii. Has technology been designed to allow or support group work?

iii. Has technology been designed to support links between different groups to develop organisational networks or systems?

iv. Has technology (and its products) been designed so as to be acceptable by society at large?

v. Has technology (and its products) been designed to take into account its impact on the environment?

2. *Designing Anthropocentric Technology: The Dual Design Approach*

The Dual Design Approach is a set of principles to ensure appropriate development of both technical and human aspects of man-machine systems (Henning and Ochterbeck, 1988). Usually project engineers tend to head for fully automated concepts (Fig. 1). Here the major part of design efforts, creativity and research is used to obtain a fully automated system. However, at a certain stage of the development it becomes obvious that certain elements of the system cannot be fully automated.

Therefore it is necessary to introduce a second approach, the working-process based design, in order to consider the human work situation as well. Contrary to the technology-based design, a working-process based design raises the issue of how to solve the problem without or with a lower level of automation or computers. This results in a concept where tasks are performed by people. Both the technology based design and the working-process based design should be used in parallel to obtain an optimum. The weaknesses and the advantages of both concepts have to be compared and analysed. It leads to concepts of man-machine systems which correspond to the demand of both, the technical processes and the process of human work. Concepts created by this approach make the best use of both the technical and the human resources. The approach may thus be considered one way to achieve economic advantages combined with designing meaningful workplaces.

This approach leads to the design of human-centred systems. It is important for this approach to integrate people into the design process at a very early stage of development. These people are the best judges as to the human-centred features of the system. In the following section, this approach is illustrated by three examples.

3. *Three examples of applying the Dual Design Approach*

3.1 CNC-machine tools for shopfloor programming: Keller (D)

One of the most important developments in production technology has been the implementation of computer systems to control machine tools, e.g lathes etc., Computerized Numerical Control (CNC). However, in many large companies, the department for production planning prepares the computer programs which control the CNC machines. They are transferred into the machines by disc. The shopfloor operators only have to start the production. This development has caused criticism and rejection among workers in many factories. They demand to be given control of the production process while making use of the advantages of CNC technology. Therefore a German software house has developed new programming tools for computer-controlled lathes.

They allow the operators on the shopfloor to create on the screen exactly the graphic contour of the workpiece they want to produce. They do not need any substantial training in computer use. They 'draw' straight lines or arcs as displayed on the keyboard. Thus the workers remain fully in charge of the production process.

There are other companies designing programs which have similar aims. One system uses manually-controlled wheels of the same kind as used on conventional lathes. But these wheels electronically transfer the data to set the machine. Thus the workers handle the machine 'manually' while taking advantage of advanced control technology.

Corresponding to the Dual Design Approach, the optimum level of automation (Fig.1) is the programming of machine tools by shopfloor personnel: one of the most important concepts of Human-Centred Production (computer assisted system). The reports of many different applications of this concept show the economic viability of the partly automated system. The programs developed by Keller are particularly useful for small and medium-sized enterprises in less industrialized countries in Europe (Keller, 1990).

3.2 A shopfloor production planning system for group work: ESPRIT 1217 (GB)

A human-centred manufacturing system is based on the principle of 'operator control'. As an example, the production aspects of it are represented by CNC. The production planning is another aspect of it. Here, too, the role of a computer is that of a tool. It represents a decision-making aid. It provides background information, and it enables the operator to estimate and, possibly, simulate various alternative options before making a decision.

In the project ESPRIT 1217 (1199) these aspects were put into practice. The aim of the project was to design as a model case, a computer-supported tool for shopfloor production planning. The manufacturing cell developed in the project produces radio connectors. They have a diameter of about 4 mm and an overall length of about 20 mm. The team working in the cell consists of four people. The tasks performed are planning of the work for the week; preparing the work and setting the machines; producing the different parts; quality and performance control; management of tools, raw materials, etc.; maintenance and minor repairs of machine tools. The main machine tools in the cell are a CNC lathe and a multi-purpose machine. Furthermore all the necessary tools etc. are on hand. Two personal computers are available to help in the management of the cell.

Corresponding to the Dual Design Approach, the optimum level of automation (Fig.1) is the design of production programming and scheduling by teams of shopfloor workers supported by decentralized computer use: group work is considered the main feature of Human-Centred Production (computer assisted system).

The model cell has been successful. The first results of its evaluation are as follows: improving due date performance by 55%; reducing lead time by about 50%; reducing working capital by about 45%. Therefore the company decided to introduce this concept of "island-based" production. Several other British companies have since implemented the program (Hancke, 1990; Rosenbrock, 1990).

3.3 Human-Centred Technology in transportation: Eurocontrol (NL)

During the late 50's, the European countries were confronted with the necessity to cooperate more closely in air traffic control. The main reason was the appearance of the jet-aircraft in civic aviation. Hence, several nations of Western Europe established Eurocontrol, in 1960. One of the tasks of Eurocontrol has been to develop and implement a system of efficient air traffic control in the upper airspace over a large area of Western Europe. The air traffic controllers at Maastricht are in charge of communicating with the aircraft flying through their area at higher flight levels. The pilots receive information on the flight level to take, besides further information on their route, weather conditions etc. The controllers follow the flight path taken, by means of radar.

Given this task, it has been the responsibility of Eurocontrol engineers to design a system which combines both people (e.g. air traffic controllers) and technology (e.g. radar, computers, displays, communication systems). In this context, the maximum use of computers may mean full automation of air traffic control - but nobody would like to fly across Europe in an aircraft with neither a pilot nor a controller in charge. On the other hand, no use of computers would mean that it would not be possible to sustain the present level of air traffic density. This dilemma characterizes the problem of designing a system of air traffic control.

Corresponding to the Dual Design Approach, the optimum level of automation (Fig.1) is characterized by computer-supported ground-ground data processing and exchange, but direct verbal ground-air communication: thus humans remain in the centre of the control processes (computer assisted system). It has been a deliberate and conscious decision of Eurocontrol to choose the middle level of automation. It may signify the limit of air traffic density because the separation of aircraft in flight must correspond to the time needed to verbally transmit information on changes of flight paths in case of emergencies (Endlich, 1990).

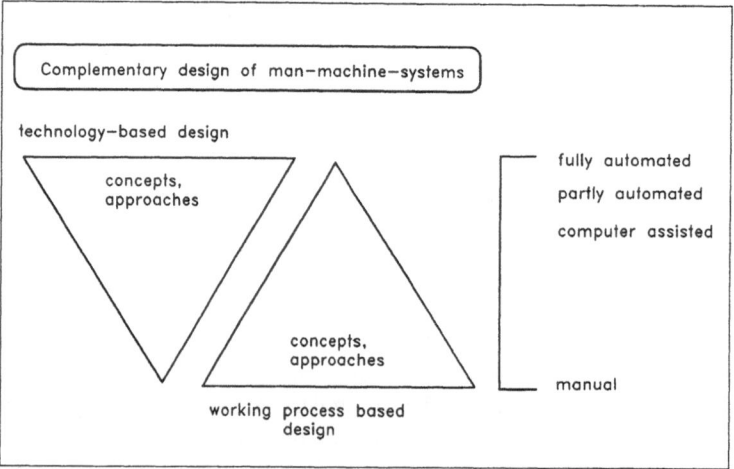

Fig. 1 The Dual Design Approach of Man-Mashine Systems

4. Conclusions

In this report, recent experiences have been described which have been gained through the process of introducing both advanced technologies and innovative approaches to work organisation. The aims have been to improve working conditions, qualifications and work efficiency, and to increase profitability. Hence it is important to have methods and strategies avaibale to evaluate technology with regard to these aims. One such method is the Dual Design Approach as described in this paper. It has been developed by one of our research teams. Another method has also been developed by our research team in order to evaluate workplaces with regard to their human-centred features. It is based on the Dual Design Approach. It uses a scale (1-5) in order to characterize a certain workplace through a set of figures. This method has been successfully tested in a series of different projects (Bohnhoff, 1991).

Many European countries, however, look at advanced technology in different terms. Three aspects may be referred to, as follows.

- A common pattern in many countries is large-scale unemployment. Millions of jobs need to be created within the next decade.
- In several European countries, working conditions are far from being "anthropocentric". Besides other problems, children are still a major part of the labour force because of lower wages.
- Some important natural resources (oil, coal, iron, copper etc.) are imported in large quantities to Europe. They are bought rather cheaply from some of the less industrialized countries outside Europe.

We may need to re-think the use of both human labour and natural resources (energy, raw materials) if we intend to change these patterns.

BIBLIOGRAPHY

Bohnhoff, A. (1991). Ein prospektiv bewertetes Identifizierungssystem. VDI, Düsseldorf

Endlich, W., H. (1990). The Maastricht U.A.C. Civil/Military ATC Environment. Presentation to the German Civil Aviation Academy, Langen (D).

Hancke, T. (1990). The ESPRIT-Project 1217 (1199). Research Report HDZ/KDI, University of Technology Aachen.

Henning, K., Ochterbeck, B. (1988). Dualer Entwurf von Mensch-Maschine-Systemen. In: P. Meyer-Dohm et al. (eds.), Der Mensch im Unternehmen Bern, Stuttgart, p. 225-245.

Keller, S. (1990). CKE + CAM, die praxiserprobte WOP Lösung, Lernfeld Betrieb, 1, p. 26-28.

Rosenbrock, H. (1990). Machines with a purpose. Oxford University Press, Oxford.

Productivity and Culture: Some Elements for the Design of a Nation's Project in the Face of the World Economic Restructuration at the End of the Twentieth Century

C. Armendariz and A. Pacheco
Professional Interdisciplinary Unity of Engineering Social and Management Sciences
National Politechnic Institute
Section of Graduate Studies and Research
Mexico D.F., Mexico

ABSTRACT

The aim of this work is to provide with some elements in order to explain and justify the possibility to elaborate a Nation's Project, departing from the reconciliation of its increasing the productivity systematically, in such a way that it will improve the people's standard of living.

1. The increase of productivity is not, in any sense, a new challenge for the Mexican Industry: but, without any doubt, the new opening and economic globalization conditions, place it in the center of the managerial dynamics. It is well known that the two determinant elements of productivity are: the human and the technological factors. If we understand on the one hand, the worker not only as an applicator of certain knowledge and abilities for the work, but also as a carrier of the culture he belongs to, with concrete interests and expectations. And on the other hand, if we depart from the knowledge that technology is not only a product of the culture in which it is generated, but also technology is a culture's bearer, then we would be able to value the importance of the relationship between them in the development of the nations.

Although the relationship between Culture and Technology has been analyzed from different points of view, that does not invalidate the need to retake it now under new circumstances. The fact is that the Economic Globalization Process (EGP), in which the world has been immersed during the last decades, and more precisely, at the end of the 20th century, has as an arrowhead the technological development, and it claims for a revision and evaluation of the correspondent impact on the cultural values of each nation. In order to make a contribution to this important issue, we must depart from the consistent conception of what Culture and Technology are, and specially, from the importance of their interaction in the development of the contries and of the whole mankind.

In this sense, we may start saying that the different and multiple ways of manifestations of human civilization, are going irremediably, towards the conformation of an one and Unique Universal Culture, which synthezises each one of the cultures. It is to be said, that we do not understand for "unique", the elementation of the rest of the cultures, and the imposition of a specific one as a substitution. That Unique Universal Culture should recover the great values generated in each of the particular culture experiences, because all of them, without exception, have something valuable to contribute to its enrichment.

This process of cultural synthesis should not be understood as neutral harmonious or conflict-free. On the contrary, so far, the human race history has been the history of exploitation of the man by the man and of a country by another country. In other words, the history has been a systematic imposition of a culture over another.

In our days, the EGP tends towards the conformation of a world-wide Unique Market (which includes not only goods and services, but capitals too) as a natural result of the world-scale development of capitalism., which requires as an strategic measure, the substitution of the Nation-State, as a fundamental macroeconomic cell, by the new concept of "Regional Block of Countries", each one with its correspondent "back-yard". In this globalizator process emerges a new international division of work, and the big transnational enterprises stablish the orientation and content, according to their particular interests of accumulation of capital. This way, we are able to assure that the EGP is the historical trend that acquires the capital internationalization process in the last years of the present century, and the integration of the so called "Regional Blocks", is nothing else but its historical concretion.

2. It is clear then, that being the EGP there is the threat of a new, or better said, the progression of a never finishing culture imposition. According to this, the subordinated cultures are obligated to elaborate and take into practice alternative plans that, avoiding the risk of isolation (it has been proven historically that isolation is a real economical, technological, political and finally a cultural suicide), would present a solid platform to face, in a better way, such cultural penetration, assimilating the central technology and culture, without loosing their own cultural identity.

From our point of view, the local cultures must be consolidated and mature in order to be able to contribute creatively to the conformation of that Unique Universal Culture. A nation which does not have clear its own origins and culture and cultural values, and consequently, does not feel identified with them, has nothing to contribute and, besides is an easy victim of cultural penetration.

In this same order of ideas, it is a mistake to talk about superior cultures, because, if culture is the result of the concrete way in which the human groups are related to their natural environment in search for survival, in a certain place and time and since reality is a synonymous of change, the diverse human culture manifestations, simply ARE NOT COMPARABLE, and they are all EQUALLY VALUABLE. But what it seems to be valid is to distinguish strong from weak economies, and one of the factors that makes that differences, maybe the main one, is the technology. Precisely, it is the economical and technological subjugation what leads irremediably to cultural imposition. We have then, that in order to measure and to be able to foresee the impact of the EGP on cultures of the Third World., it is essential to define and characterize two concepts: Culture and Technology.

From the multiple definitions of culture, we chose the most adequate for our work, which is the classic concept precised by Edward Burnett Tylor: "Culture... is that complex whole which includes knowledge, belief, art, morals, law, custom, and any other capabilities and habits acquired by man as a member of society". That is, culture includes all activities of human life. Every society has its particular and distinctive cultural expression, for it is its principal agent to preserve stability and to secure its continuity, Culture is the most powerful force, that exits in society, for its great influence in each one of its members. Culture is many times, most powerful than survival instinct.

Propagation of culture happens all the time, and it has always happened in every place and from generation to generation. It is "contagious" and it is too easy to spread among people of similar customs, from one region to another, whenever this culture offers certain advantages, usefulness or pleasure, to be accepted in a natural way.

Nevertheless, there is another process, the "transculturation", that consists in the spreading or influence of the cultural characteristics of a society, when it comes in touch with another, that is much less evolutioned. This phenomenon may happen among people that have nothing or very little in common. This is, precisely, the case of conquest and colonizations. The conqueror's culture is imposed over the conquered and, most of the time, this is done in a very violent and cruel way. Such as it happened during the conquest of Mexico by Spain in the 16th century, and which has been described by G. Bonfil as "the most brutal demographic catastrophe that history has ever known".

Transculturation, nevertheless, is given in other ways more subtle and much less violent, disguised many times, as progress or modernization. This kind of transculturation is less drastic and inhuman imposition, for its effect is imperceptible, and due to this, it does not find neither opposition nor obstacles in its path. One of the most common penetration ways is through technology, for it is the best decoy that exist to entice the countries with a low economic level.

As well as the term Culture, Technology has many ways to be defined, that goes from the specific to the general. For our essay, we used Arturo Pacheco's definition: "Technology is all that helps the interaction between MAN and NATURE in any production process of satisfactors, and its objective is to increase the power of human work". In that sense we can assure, that technology is as old as manking itself, and that its evolution goes hand in hand with the evolution of man. Normally, it is associated with the process of generating wealth. Nevertheless, technology does not always benefit the nations that import it, and not only that, but many times it carries along more problems than those it was supported to solve. That is why technology should be used in all those tasks in which it can help man to perform in a better and faster way his work, freeing him to do other important activities according to his inclinations and interests. But technology should not substitute all human efforts, because if that happens, it would be depriving man from one of the greatest satisfactions and enjoyments of life which is to endeavour to do things by himself, trying to reach personal achievment. That is to say that technology should go as far as the human nature permits it, and it should serve lastly, to dignify the human life.

It is clear than, that the EGP for its own nature, affects all the nations of the world. But in a very significant way and with a much greater impetus, impacts the nations with dependent economies. This is the case of the Latin-American countries which occasionally, are forced to sacrifice certain cultural values in order to obtain a supposed modernization that everybody hopes it will come to raise people's standard of living. By saying this, we do not mean, that the trend for the internationalization of economy, finances, politics, science, technology, etc., has only a negative side. This process, of course, has also a positive side. Nevertheless, the price that all dependent countries pay, to have certain economic advantages, it is very high; so high, that sooner or later, may derive in the loss of their national identity and, perhaps, even their self-determination and sovereignty.

For Mexico, specifically, the signing of the Free Trade Agreement (FTA) in 1992 with the United States and Canada, is nothing else but the concretion of the EGP. And the disadvantages that Mexico might have to endure, will be same or pretty similar to those that the rest of Latin-American countries are suffering already: more exploitation of manpower; unfair competition in the production of goods; damage of the environment because of more industrial residues; overcrowding and increasing of the prices of public utilities as well as housing in the places where companies stablish; changes in the way of thinking and understanding life (ideology); more technological dependence, etc. But, besides all these disadvantages that Mexico would share with the rest of Latin-American countries, there are many more because of its closeness to United States.

In a very general way, we will mention only three main cultural risks that Mexico is taking in the signing of the FTA:

a)   Craft traditions loss.

b)   Control loss of Cultural Industry (press, radio, cinema, and television).

c)   Ideologic penetration in the Mexican Educational System.

These three cultural dangers, which effects are difficult to evaluate completly at the moment, are only a sample of those Mexico will have to face in the near future. There are ore, a lot more, but these are just some of the ones that could affect directly the roots of our social structure through our traditions, our identity and our ideology as a nation. Below, we will discuss briefly the risks mentioned above:

a) Mexico has always been distinguished for its great vocation for craftsmanship. The little learns from his parents the skill to which he is devoted and lives from most of the time. He passes it on to his children and so on, from generation to generation. The tradition is sacred, almost as if it were a religion.

What is going to happen with all these craftsmen when the FTA comes in full to our country? We are anticipating that all or many of those artists, because of the need they have to increase their income, will leave the family's workshop to go to work, for example, in a transnational assembly plant, to which the only important thing is to earn the maximun possible. Maybe in the plant, the craftsmen will find himself performing an automation task, where there is no room for a creative spirit. Little by little he will forget, along with his skill, his roots and his own Mexican identity. He will become a different man: empty and without artistic aspirations.

b) The second risk we have mentioned is the one related to the Culture Industry of the country: press, radio, cinema and television. It is useless to some extent, to point out what this issue could mean to us. We all know the powerfull weapon that could be the handling of Mass Media by foreign interests. Through it, it is possible to convince, dissuade, cheat, manipulate and change the mentality of a whole country, specially, if it is not prepared neither psychological nor intellectually for the suden attack.

It is important to underline that during the signing of the FTA between United States and Canada in 1988, the matter of the Canadian Cultural Industry was not even discussed. because this country stated very clearly its opposition. Nevertheless, in the signing of the tripartite pact of these two countries with Mexican Cultural Industry is already being considered for discussion.

c) Undoubtedly the main mechanism that a State has to reinforce the national identity, and to preserve the cultural values which are the reason for a nation to exist, is the formal scholastic education. Therefore, it is very dangerous to give-up educational spaces, even if they small, in favour of other culture's interests. The EGP and, the FTA in particular, are creating this problem for the Mexican Educational Sytem (MES), because it exists the risk of adoping schemes and functioning parameters similar to those of our neighbour of the North, not only from the technical point of view, but from the ideological too.

Nowadays in our country, such ideological turn is taking place slowly but firmly, at Bachelor as well as at Graduate levels, through the implantation of evaluation mechanisms similar to those of North America, which give priority to the technical over the social and the humanistic parameters. In practice, this means that our future technicans, engineers, economists, managers and finally, our gouvernors, will have a professional profile that will be cleared distorted and with serious deficiencies to be able to understand and to defend our cultural identity and our sovereignty. At public Elementary and High School levels something similar is happening, but hidden under the name of Educational Reformation, which is based in a purely pragmatic conception of the educational process.

Finally, and facing this globalization situation so aggressive for local cultures, we will expose next, some proposals trying to soften the harshness of the risks mentioned above, so the integration of Mexico to the EGP, permits the country to assimilate the advantages of the technological advance, without having necessarily to change its cultural values.

Above all, it is necesairy to understand that the cultural phenomenon only represents one of the edges in the social reproduction processes. That is, of the processes that permit societies to secure objective and subjective conditions, essential to their economic and social reproduction. In this sense, a strategy that pretends to safeguard the identifity and the cultural values of a country, will have to be inserted in a larger scope strategy, in a Nation's Project. Due to this, we suggest that the rest of the following proposals, should not be considered isolated, but instead, they should be in any case, inserted in our Nation's Project, which, by the way, has not been yet elaborated.

Regarding to Mexican craft traditions, we are proposing three things to the government: a) supporting our craftsmen financially and with comercialization channels without the participation of intermediaries; b) stimulate the formation of craftsmen cooperative enterprises , and c) fortify and expand the teaching center of our craft traditions. These three measures, we consider, would contribute to make our craft enterprises more competitive in the new world marketing conditions.

Concerning the possibilitiy of leaving the Cultural Industry (mass media) of the country in foreign's hands, our proposal is a categorical NO to the possibility of discussing the issue during the negotiation of the FTA. Although there are the necessary controling regulations in Mexico for this industry, these are not being applied rigorously, specially concerning the contribution of the mass media to lift and stimulate the patriotic spirit. For this reason, we propose that the national legislation of the Cultural Industry should be revised, specificallay in what concerns to the "culture's comercialization", so the regulation would secure the preservation of our cultural values.

In reference to the underground transculturation in the MES the answer is in the hands of the educational institutions at all levels. It is urgent that the educational community seriously analyzes the negative consequences that would have in our national identity, to copy without questioning the Northamerican educational model. And, bearing this in mind, the Mexican educational community that permit the formation of highly qualified citizens for the job, but with a solid national conscience of social solidarity.

BIBLIOGRAPHY

Quotation from "The Concept and Components of Culture" (1988). Encycolopaedia Britanica (Macropaedia), USA, Vol. 16, p. 874.

Definition in Pequeno Larousse Ilustrado (1982), Editiones Larousse, Mexico, p. 1016

Bonfil, G. (1990). Mexico profundo una civilizacion negada. Editorial Grijalbo, Mexico, p. 127.

Pacheco, A. (1990). Guia para la Instalacion del Programa Permanente de Mejoramiento de la Productividad (PPMF). Cuaderno de Investigacion No. 5, Editorial IPN-UPIICSA, Mexico, p. 5.

Training of Human Resources for Automation Design in Developing Countries

L. del Re and W. Ipanaqué
Institute of Automation
ETH - Zürich
Zürich, Switzerland

ABSTRACT

Automation reduces the number and necessary qualification of most workforce in plant operation and so allows to enthance the quality of products also in countries without enough training workers. However, by reducing the need and importance of skilled workers it reduces also their action field and does not contribute to spread knowledge.The higher requirements for the automation design may even deepen the dependance on foreign know-how, thereby offsetting the positive effects on the long run.

In order to avoid this, enough trained human resources at the design level are necessary. Countries with low engineering costs become then even able to export automation, as shown by examples in India and Chile.

The necessary technical training must be adequated to the needs of developing countries, coping with the lack of an industry able to introduce young graduates to the practical side of engineering. Furthermore, in a developing country even a beginner can seldom count on experienced collegues and my have to start by discovering the needs, what requires a certain professional maturity.

The proposed paper relates a field experience in a small high school in Northern Peru, where a human resource developing program was started in 1981 with the goal to constitute a long-term training unit in the field of applied electronics and small scale automation. The idea was to build a group with the task to offer services and courses to the local industry and to train engineers. In order to achieve a durable result, the group was expected to find its social function in a few years, to become at least partly self-financing, to develop a "recursive" recruitment and training scheme and to implement an internal continuing scheme.

The whole project was organized with three overlapping steps, in order first to motivate, then to support the acquisition of technical skills and at last to promote the insertion of the group in the industrial context.

The group was actually built, developed constantly both technical skills and self-confidence. As a concsequence, after a few years it begun to fulfill the assigned task, first by an up-to-date teaching in the field, later by classes for industry engineers and then by consulting and project activities. Although more steps are still needed to improve the level and the durability of the results obtained, first reports confirm that positive effects of this ten-years long effort are beginning to be felt in the local industry.

1.   *The impact of automation on developing countries*

The increasing use of automated equipment has a social impact at least in two years. On one side, the quality of the products is enhanced and the highter efficiency makes these goods available to a wider part of the population for less money. Unhealthy or dangerous jobs can be reduced or even eliminated.

On the other side, the composition of workforce is changed. Earlier fears that automation would increase unemployment rates have not proven true, perhaps because the pretentions of customers have increased as quickly as productivity. However, some qualifications are no longer required, among them many tasks for skilled workers, what has a social price in any society. As at the same time more design hours of engineers are required, the relative importance of the plant design phase and of the operation have been changing.

In a developing country, automation has still another effect, as it may interfere with technology transfer, generally thougt to be a necessary component of any effort to reduce living standard differences between developing and industrialized countries. This transfer does not happen simply by importing advanced technology plants and having some technicians to train the operation personnel for a while (El-Hares and Tayel, 1980). Even if the necessary technical informations are transmitted, their scope is very limited and they do not contribute to the wished shift from dependance to interdependance (cfr. Wagner et. al., 1983). A developing country does not need only to solve some specific problem, for which import may be the right solution, but also to acquire the corresponding skills, i.e. the cultural patterns that give the ability to define, analyze and solve other similar problems using the available tools. This implies merging some elements of a foreign culture into its own, a process happening anyway.

This merging include acquiring more capabilities or simply mean a shift in consumer habits towards even less understandable goods, increasing the know-how gap. Which kind of merging actually takes place depends mainly on the availability of human resources able to act as a bridge. As automation reduces the action field of skilled workers, this task must be assumed to a greater extent by the higher qualification level, the design engineers. These, however, have a strong propensity to become alienated from their own culture rather than to act as a bridge.

Automation is the solution designed for some problems of developed countries, and may prove useful for many problems ot developing countries too. But automation reduces the importance of some assets of developing countries like cheap labour and makes the treshold know-how level required for up-to-date - i.e. efficient - production higher. This means that development has to rely even more on education than earlier. This education must be adequated to the needs of the developing countries: this does not mean that it should be incomplete but that knowledge and skills in the former defined sense should be learnt together, getting the student used to recognize the possible uses of the technology in this environment.

2. *The project environment and structure*

Similar considerations motivated in 1981 the establishment of a unit of teaching and applied research in electronics and control, with the goals

1) to train young engineers to the use of the technologies available on the market and

2) to assist local industry to the same purposes, as well as to help solving specific problems.

The project was centered in Piura, a town of about 300,000 habitants in Northern Peru, about 1000 km by road from Lima. Piura is considered since aq long time a potenital 'agroindustrial´ development pole due to its rich agriculture potentiality, its main large industries being food and cotton processing. The project was developed together with the University of Piura, a private but open to everybody High School, financed to a large extent by voluntary industry contributions, and with a long tradition of international cooperation. At the beginning of the project, the University of Piura offered courses leading to the Industrial Engineer degree in six years. This career included a few administrative courses and a wide overview of civil, electrical, mechanical and chemical engineering. In order to obtain the official 'Ingeniero´ degree, the students hat do prepare a thesis, which could take one year and offered them the opportunity to deepen their knowledge in one of the fields. Some students ware then offered a position at the University, and were charged with some course. Very few experienced professors were available, mainly with administrative and seldom with technical experience. Details of the curriculum are described in (Azzaroli and Bonavia, 1987). An important advantage of the curricilum of this University is the interdisciplinarity, so that the 'system thinking´ was, at least theoretically, easier to introduce.

Some problems were typical of the sector at the beginning of the project: high turnover, low motivation, incomplete technical background and few opportunities and no habit to update their knowledge through courses or written information. The human resource training program to be conceived must lead to establish a working group with some essential features:

a) the group should reach a stable stationary level - i.e. natural turnover should mean no danger to continuity or level

b) the group should be oriented to the actual needs of the region and

c) the organization of the group hat to be such to allow a continuous level increase also outside of cooperation projects.

Many approaches may be taken to reach these goals. In our case, it was decided to use local students and to complete their training in the specific field, instead of 'importing' electronics engineers from Lima or abroad, to use state-of-the-art technologies, to have the technical and the professional training contered at Piura, using stages or courses abroad as a complement, but not as a substitution of the local training. The program was designed according to a three steps approach:

a) a motivation step, needed to attract students of the last cycles and to keep some of them interested while filling the gaps of their technical know-how (with foreign personnel on place),

b) a step of induced growth, intended to stimulate them form outside and support their efforts to improve the technical level and

c) a 'landing' step, during which both a gradual increase of technical activities for the local industry and a change from the cooperation based assistance to a more or less normal inter-university cooperation should take place.

The first step consisted in awakening the interest of young students to the field by organinzing simple introductory works, like constructing a digital watch. This helped to have a small group interested in learning and applying the methods they were being proposed. A slow staircase of more complex problems was designed, so to increase both level and motivation. The 'elder' students were induced to assist or even assume the responsibility of the training of younger ones while continuing their own. At the end of this phase, foreign personnel left.

The following step, from 1983 to 1987, aimed at increasing the autonomy and know-how level of the group, while keeping the energy diverted to everyday problems as low as possible, for instance by having external consultors acting as purchase - and financing - office. During this phase, a certain degree of technical dependance remained, and suggestions for new activities had to come mainly from outside. The third step, from 1987 to now, has been characterized by an autonomous activity, helped, integrated or corrected by the external help, but led frem the local group. In this phase, first important contacts and works for the industry have happened.

An essential point of the project was the 'seed' idea used in the second step of the project. It is often regretted that technical cooperation does not simply offer the means to solve the underdevelopment problems, but tries to give also the solutions. On the other side, many wrong decisions are taken in developing countries due to the lack of information or professional maturity. In order to reduce both problems, many different small projects were "proposed" in different fields instead of fixing the exact working field, hoping that some of them would find the adequate environment and prove useful. Between 1984 and 1987 courses on different aspects of industrial controls were held in Zurich and Piura, and "examples" were taken to Peru, ranging from gasoline motor ignition observation to pneumatical drives. Small test rigs were built, asking the Peruvian partners to start small jobs - "tesinas" - on them. Some of the activities proposed were abandoned very soon, others are still mor or less at the level they were introduced. Others, notably PLC and vibration monitoring, experienced a great success: in both cases, the key factor was the combination of a minimum equipment, the right person and a true application in the region.

3. Conclusions: the impact on the local industry and the validity of the scheme

In order to judge the validity of the scheme, an assessment of the practical results as well as a comparison with alternative approaches are needed. As stated before, neither the continuity of the group activity nor its ability to continue enhancing the technical level seem to be in danger. The impact on the industrial world of the region is actually just beginning, but is already taking shape in three forms:

1) Througt the graduated students working now in the industry. As they are used to an "active attitude" toward modern technology, they are able both to recognize some inefficient use of machinery and to consider alternatives. First feedback of this kind is arriving, although much a greater impact is expected once they have had time to climb to positions of higher responsibility - the first "new" generations left the University 1986. In case of former employees, who generally get immediately some greater responsibility, the feedback is quite stronger and has led for instance to automation contracts for some equipment of oil platforms.

2) Directly, through courses for engineers in the industry, like a local version of the Zurich course on measurement techniques of PLC automation. These courses provide usually an opportunity to discuss actual problems of the local industry, and reduce the passivity of many local engineers or technicians towards technology imports.

3) Some of these discussions have led to cooperations, that have solved long standing problems, like the automation of the fire extinguishing system on an oil loading dock. usually, these cooperations have been possible with public enterprises, like Petroperu, Petromar or the City Council of Piura. However, the number of plants that could profit from such cooperations is very high, so that adequate publicity on the practical implementation of locally engineered automation on local plants is being prepared in order to sensitize other firms. So far, both technical training level of the students as well as the number and magnitude of industry-related activities are increasing.

Fig. 1 gives an idea on the different increase or decrease rates according to the phases of the project.

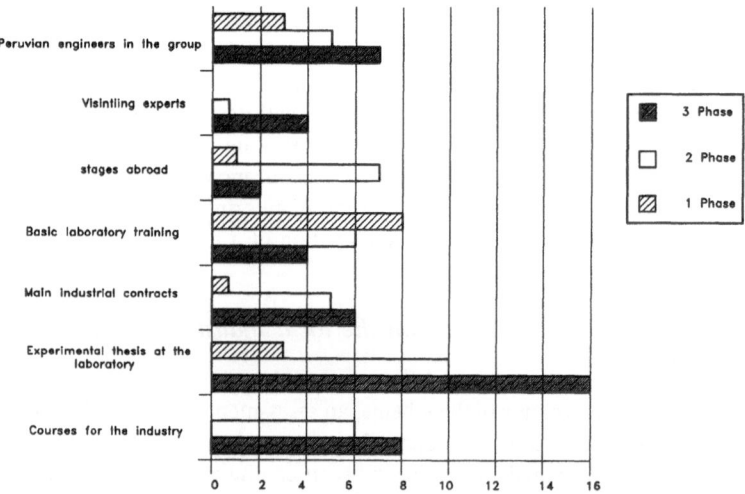

Fig. 1. Overview of some aspects of the developement of the project

It has been an open discussion whether the gradual approach used here was an efficient one. Clearly, introducing graduated electronics engineers would have presented many technical advantages. In other groups at the same University, the other approach has been taken, but it is still too early to make comparisions. However, some conclusions may be already drawn:

a) the human resource training has led to a group that is assuming a bridge function without being alienated from the local culture. The approach used seems to be leading to 'maturity´, defined as the readiness and ability to do something without being taught. Using locally available personnel and training at an adequated age proved sensible.

b) The price is the slow pace of increase in technical know-how.

During the program it was tried to create new groups in other fields by increasing the dimensions of the first groups and then splitting it. This approach has worked only it the two new groups worked in very near fields. This was surprising, as the very general background of the Piura industrial engineers should have allowed the step without much difficulties.

The results of the program may be considered quite positive, comparing with the classical problems of many University cooperation projects and keeping in mind that the region of Piura has been a very tough environment - extremely damaged by the alluvions of 1983 and experiencing in the last years a deep recession as the whole country. We believe that this success must be attributed also to the combination of 'human factors´

that started and appoved it. However, the greater merit may lie in the step-by-step approach, that by simultanous increases in competence, responsibility and autonomy allows the local grow in full consciousness of the bridge function and may therefore prove, on the middle term, a very fruitful way to provide developing countries with the necessary human resources to make of automation a contribution and not an obstacle to development.

4.   *Acknowledgments*

The actual implementation of this project has followed the guidelines exposed, but the single steps during the years have been adapted to the actual developments of the Peruvian group and to the available resources. The second and third step consisted of different actions, most of them executed by the Istituto per la Cooperazione Universitaria in Rome with the financial support of the Italian Government, by the ETH of Zurich, with the a co-financing of the Swiss Directroarte for Development Cooperation, or by both.

BIBLIOGRAPHY:

Azzaroli, P., Bonavia, M. (1987). La formazione degli ingegneri presso l'Universita' di Piura in Peru, L'Elettotecnica, Vol. 74, No. 2, Febbraio, p. 181 - 188

El Hares, H., Tayel, F. (1980). Optimal implementation of process control in developing countries. Proceedings of the 3rd IFAC Symposium on "System Approach for Development", Rabat, Marocco, p. 489 - 491

Elizonde, J. (1980). Education and its aims. Proceedings of the 3rd IFAC Symposium on "System Approach for Development", Rabat, Marocco, p. 403 - 409

Wagner, N., Kaiser, M., Beindiek, F. (1983). Ökonomie der Entwicklungsländer, Gustav Fischer Verlag, Stuttgart, FRG

Socio-Economic Validation of a Modular Combinative and Non-hierarchical Approach of Integrated Production Systems

C. Everaere and C. Mahieu
Federal Institute for Research in Economie and Industrial Societies
Programme 3.I.E.
Lille, France

ABSTRACT

Confronted with technological aspects of "CIM", a research, gathering engineers, economists and sociologists, based on information and decision-making systems in an unpredictable and unstable environment, focusing on social and economic aspects.

1. Facing a Reactivity Context

Studying the necessairy conditions for the evolvement of manufactoring systems and their context (M.R.T., 1989) has emphasized a reactivity principle in order to characterize an environment dominated ba a loss of stability and predictability. As a result, organizations have to look for alternatives to a planned, stable, rigid and sequential running; which are prevailing princips in the functional or divisional traditional organization (by product, area, market...).

In this situation, it seems fundamental not to seek a rigid and definitive structure which could only apply to a limited number of cases. But, inversely, we should conceive an adaptable flexible and thus reactive structure, i.e. an open system without a priori which favour information flows instead of too deterministic material flows.

2. A Model of Global Integration Through Data

Data integration focuses on the particular care given to data circulation from the environment into the firm, so as to understand as quickly as possible the nature of the external and internal disturbancees which affect the daily running of the firm, in a global vision being attentive to the great interdependance of functions.

Hence the importance of computer systems to link the various components of the global system whatever their location, but above all the unavoidable postulate for an information system, to which any body might access, being sure that information is the same for anyone and its using do not affect its consistency.

This is a first condition of reactivity, which can be compared with an "informational vigilance" enabling to know any aleas and to take them into account in the running and regulating of the production system. That is why the interest is drawn on "data-integration" : shared data accessibility, circulation, use, memorization devices.

3.   The Integration in Question

This conception of integration leads us to debate on the possible misunderstandings about this very used but rarely defined concept. At a time when the integration concept (e.g. "C.I.M.") is gaining ground as a necessary alternative to a rigid, compartmentalized and algorithmic organizational running, we have noticed that the integration concept bears some contradictions. From research about the implementation of automated, flexible and in some way "integrated" production systems; and with studies about the principles of organization and production management in Japan, we confront "integration" meaning of getting closer, compact, concentrate, agglomerate, to that of join, connect, link and coordinate.

As fare as the conception of industrial production systems, made of various structural components in terms of functionality, use conditions and specificities, is concerned, we favour "physiological" meaning of integration and we insist on the interest of a shared and distributed system rather than a concentrated and compact one. Within this conception of integration, the interest of networks linkage systems is obvious.

4.   A Flat and Combinative System

An other consequence of reactivity is to question the places and methods of disturbances processing: the traditional pyramidal hierarchical system is articulated around a centralization of data processing means, of problem-solving and decision-making. If, owing to an unstable environment, the structure continuously warps, there are too many and complicated problems to be solved "on the top" in their entirety and in proper time.

Besides the reducing and simplifying character of a functional hierarchical system, this one leads to a principle of breakdown into functions and successive subfunctions which are supposed to intervene sequentially.

This creates partitionings which generate retaining and opacity in information flows. If an information is not available inside the firm, its user must either guess it, either negociate it or recreate it with huge error risks and, in any case, non value-added works.

That is the reason why it is propounded here to resort to a "flat" system made of functionalities (modules) which should be "activated" according to production requirements and to evolving availabilities.

5. A Proposal of Flat and Modular Architecture of the Firm

Thus we suggest to define a model of the firm in its stage of structuration and organization, that respects the physiological meaning, put forward to apply to the integration concept, and which is suitable for flexible organization requirements.

The structural components of this model are called "modules" so as to convey the idea that the firm may get organizational quality and variety thanks to a combinative principle which coordinates the functionalities - modules - in a flexible and heuristic way, in order to be able to react to external (market) and internal (hazards due to machine and human failings) events. Each module is defined according to what it makes and to the conditions in which the acitivity or functionality is achieved, and not according to traditional big entities like the workshop or the function.

In this combinatory approach, it is fundamental for each module to own the means requested for an autonomous working, in particular human and/or computer intelligence in order to manage the processing gears assigned to it, but also to obtain and supply the information required or pruduced. Processing (on material or informational flow), intelligence and communication are the three indispensable components of any module without any of them, module cannot exist and work.

In such an organization, there is no more hierarchy but, on the contrary, an active and flexible partnership between modules. Each one is responsible for an expertise area and is independant and sovereign in the internal management of its local intervention area.

6. Competence and Stability

Whatever flexible productive combinings are, the modules remain rather stable: the flexibility results more from organizational adaptability than from the structural components themselves, which might have a limited functional elasticity, due to the pressure of the growing complexity of industrial production requirements.

Let us notice the paradoxe due to the connection between the research for a flexible functioning suitable to the disturbances, and the fragilization of the system as a result, which leads to stabilize people on well defined functional areas (modules).

This is linked to the acknowledgement of the unavoidable character of human intervention, should it be for the sole reason that this one is more and more necessairy owning to the boundaries of automatism themselves. The availability and performance of the system depend on those of the individuals who make it up.

This implies the possibility for the operator to acquire real skills but also, owing to the cognitive bourdaries of individuals, the necessity of a relative stability in the job station (module). The aim is to overcome the complexity generated by the "product-equipment" couple at short term and at longer term to define the axis of improvement and evolvement of the modules.

Even if the job rotation, which can be assimilated to a certain extent (Dadoy, 1990) to a form of polyfunctionality, allows a greater flexibility in the assignment of people; as a drawback, it reduces the competence product-equipment to assure the availability and performance of the system.

But how ensuring therefore a general sensibility of the actors to the overall system, if not by a preliminary functional mobility enabling the latter once "stabilized" to understand their contributing in the whole process, and to be able to communicate with other modules on common understanding areas.

We advocate thus of "vertical" responsibility and management attached to well defined and limited complexity fields. Taking in charge a module means a deepened (and not widened) decisional capacity in a dynamic way, both "spatial" (initial functional mobility) and "temporal" (possible evolvement of the modular intervention fields). The notion of learning is at the core of this dynamic process.

7. Reactivity and Autonomy

Acknowledging a local decision-making capacity, and thus autonomy and sovereignity in the taking charge and management of local complexity fields to provide reactivity, raise another paradox. Indeed, given up an "authoritative" system, leads to a type of regulation based on negociation, transaction and retroaction between a larger number of decisions-makers.

These ones can be more difficult to convince, and the procedures of explaining and convincing so as to get consensus about decisions could endanger reactivity itself, because of the time of decision making.

We can solve this paradox by considering the global accomplishment time of a project which includes both the phase of decision making comprising a process of dialogue, conviction, anticipation and justification, and a phase of execution.

We set out the postulate, illustrate by the following figure, that devoting more time to the decision making stage can have benefic effects on the overall accomplishment time of the project, because the anticipation of problems by the preliminary emergence of objective difficulties or disagreements, at the stage of conjectures level, will avoid a waste of time and resources if these problems appear at the stage of execution.

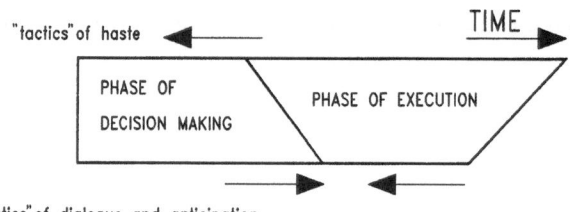

8. *An "Embedded" Dynamic of Evolvement and Development*

How making a flexible organization evolve without requiring heavy reconfiguration? It is precisely at this level that a modular structure should prove its efficiency. The design conditions and the implementation of such a structure make the organizational development as it were "embedded".

The points of not the type of socio-technical step accompanying the modification of the functional structure, but the elaboration of an organizational model likely to make the operating conditions vary continuously.

This dynamic running implies a reconversion of the organization by reconsidering continuously the connections between modules. The evolvement of the organization is constitutive of the modular structure which enables an unstabliliy of the connections between modules, i.e. an organizational reactivity. At the same time, it grants stability of the modules through the integration plan of the firm, which determines the strategic objectives of the firm and the means (internal or external functional entities, human and technological resources...) to achieve them.

The reconfiguration of the organization is not the leading of a transition between two successive states of the functional organization but a very dimension of a modular structure. Then the notion of transition takes a new sense.

It is not a question of progressive developments from a planned change of structure, readjusting connections defined according to an unique expected use.

Thus we pass from a conception in which the "socio-technical" and "participative" step is added to an important reconversion/reconfiguration, to a continuous step which can do without such reconfiguration by objectivizing and conceiving an organizational "plasticity".

The socio-technical step is no longer heavy (gathering people specifically,) long (because external to the temporality of the productive activities) and expensive in deep reconfigurations supported with normative and algorithmic methodologies. This step has become integrated to the concrete implementation of a modular system made of autonomous intelligent entities with appraisal and management capacities. The socio-technical step is at the core of the modular system.

9. *As a Conclusion....*

Behind the specifically technological transformation are emerging changes which affect not only the industrial organization model but the "invisible technology" (Berry, 1983) too, that are the management tools.

An "embedded" dynamics of structural evolution, the only one liable to make the most of the potentials in term of industrial efficiency brought by the advanced information and communication technologies, implies that new appraisal economic tools have to be (thought out?), clarified and appropriated by any actor in the firm.

BIBLIOGRAPHY

Berry, M. (1983). Une technologie invisible? L´impact des instruments de gestion sur l'èvolution des systèmes humains. C.R.G.

Dadoy, M. (1990). La polyvalence et l'analyse du travail. Collection des Etudes CEREQ No. 54, mars 1990.

Freyssenet, M., Thenard, J-C. (1988). Choix d'automatisation, efficacité productive et contenu du travail. Cahiers du GIP Mutations Industrielles, avril 1988.

Jameux, C. (1989). Pouvoir et organisation face aux nouvelles technologies de l'information, Revue Francaise de Gestion, mars-avril-mai 1989.

MRT. (1989). Programme de Recherches sur l'Intégration pas les Données dans l'Entreprise: P. Cohendet, A. Colin, J-P. Durand, Ch. Everaere, P. Llerena, Ch. Mahieu, "L'Entreprise face à l'integration". Rapport MRT, 1989.

Rockart, J.F., Short, J.E. (1989). I.T. in the 1990s: Managing organizational interdependance. Sloan Management Review, winter 1989.

Computer Aided Creative Process in Fine Arts

D. Sirbu and P. Kopacek
Romanian Ministry of Youth and Sports
Scientific Interdisciplinary Programme
Bucharest, Romania
Scientific Academy of Lower Austria
Krems, Austria

The artist belongs to his epoch. The general level of the human knowledge will influence his artistic view of the world and environment. The techniques, methodes, materials and tools used by artist in the creative process are dependent on the contemporary civilization achievments.

The computer is perhaps the most significant representative of our material civilization. It is a completely different tool in comparison with all those used by artist during the whole history of art because it may be used both as simple execution tool and, in the same time, as an intelligent agent capable to assist and cooperate with the artist in the creative artistic process.

To be able to use the computer in the creative artistic process, the artist will be obliged to modify his view on his manuality, the so called "savoir-faire". In the same time, using computer will not double the traditional creative artistic process. In fact it will be an interaction between the artist and an intelligent instrument capable of completely modifying all the mechanisms of the act of art.

The paper analyses the way in which the process of artistic creation changes its content in the context of computer introduction in fine arts. The drawbacks in computer education of the artists and fine students are considered. The specific of artist and formation is outlined. The artist refusal to accept the computer as a tool is analysed. In the same time, the paper discussed the modifying statute of the artistic work due to the continuous reducing importance of the artist "savoir-faire" and in the same time due to the reducing degree of oneness in the artistic process.

A systemic view of the artistic creative process is proposed. The specific of the fine arts creative process is analysed in order to understand the data processing mechanisms intervening in the act of art.

The artist is described as an expert which have to take data based decisions in the creative process. The artist's rules, knowledge and data based creation process is analyzed. This is dependent on the social and historic context in which he was formed, on his genetic structure and on the currently accepted systems of norms.

To be able to analyze the psychology of the artistic creation one has to start with the problem of characterizing various systems of human knowledge, beliefs and concepts establishing how these are organized and which are the principles that underlie them. All these elements are needed in order to analyze the nature and techniques of artist's data based decision during the creative process. A classification is carried out to establish the input data and the unexpected elements which can be interpreted as noises of the process. Some general rules which are taken into account by an artist during his work elaboration are analyzed in order to establish the reference elements within the systemic view upon the artistic creation process.

The analysis of a work of art as the result of the artistic creative process assumes the understanding of the artist's intentions and of his psychological motivations. Chomsky' said that "the study of language structure reveals properties of mind that underlie the exercise of human mental capacities in normal activities such as the use of language in the ordinary free and creative fashion". The grammers in mathematics were introduced on the basis of language structures study. Taking into account their importance in understanding the human mind mechanisms and the strong subjective character of art creation, grammers are proposed as tools for computer approaching of the artistic domain.

BIBLIOGRAPHY

Cornock, S., Edmonds, E. (1973). The Creative Process where the Artist is Amplified or Superseded by the Computer. Leonardo, Vol. 6, p. 11-16, Pergamon Press.

Tijus, C.A. (1988). Cognitive Process in Artistic Creation: Toward the Realization of a Creative Machine. Leonardo, Vol. 21, No. 2, ,p.167-172.

Gsolpointner, H. (1991). Art and Design. Second International Workshop on "Computer Aided Systems Theory, EUROCAST´91", April, 1991 Krems, Austria.

Chomsky, N. (1972). Language and Mind. Harcourt Brace Jovanovich Inc., Second Edition

Passeron, R. (1974). L'oeuvre picturale et les fonctions de l'apparence. Librairie Philosophique J. Vrin.

Da Vinci, L. (1971). Treatise on Painting. Meridiane Publishing House (in Romania).

A Tutorial on Computers in Visual Arts

D. Sirbu, V. Sirbu and P. Kopacek
Romanian Ministry of Youth and Sports
Scientific Interdisciplinary Programme
Bucharest, Romania
Scientific Academy of Lower Austria
Krems, Austria

Extending our previous interests in computer aided education (Kopacek, 1987) we are currently concerned with the study of the interdisciplinary methods in artistic education.

This paper discusses the current status of computer technology in visual arts. The authors briefly review the history of the computer art. The main directions and results in the introduction of computers both as a tool and as a creative intelligent agent in the contemporary art are considered.

According to Franke (Franke, 1971) the short history of computers in art is the result of two distinct lines of development: computational graphics and machine-based art. For the first time computer graphics were generated by means of analog systems by Ben Laposky (1952). Digital graphical words were published simultaneously in Germany and USA (1963) by Nake, Nees and Noll. The first figurative computer art work was displayed by Charles Csuri (1967). Since then more and more artists and computer scientists were concerned with this interdisciplinary creative technique.

The technological drawbacks and the ideological content of art in the ex-socialist countries explains the lack of significant results in computer art in Eastern Europe.

The paper contains a survey of the current works on different topics in computer art; VLSI, workstations and peripheric equipments, graphics standards, image acquisition, digitizing, processing and generation techniques, human-computer interfaces, computational geometry, etc.

A special attention is accorded to the interpretation and modeling of the artistic creative process. Increased understanding of using color, texture, line and form is considered to be fundamental in the establishment of new trends in the computer art creation process. Different aspects of the creative process were already presented (Tijus, 1988; Cornock and Edmonds, 1973). A systemic interpretation of the artistic creation process has been suggested, by us (Sirbu and Kopacek, 1991). Based on this proposal, the requirements imposed in the development of an integrated system dedicated to the interdisciplinary education of the arts students are discussed.

An extended analysis of the possibilities of introducing computers in different visual arts vis. painting, sculpture, graphics, illustration, monumental arts, ceramics, fabrics, photography, movies, animation, etc. is presented. A special attention is accorded to the computer aided education in art. Other domains of applying computers in art are considered: restoration and conservation of art works, art history, art aestetics, art critics, forgery recognition, etc.

According to Franke (Franke, 1985) "computer art will develop into a new field of aestetically - oriented activity which can neither be classified as part of existing classical branches of art, nor must it to be recognized as art at all. Thus a new profession could emerge as was the case with photography and cinematography". Considering the dramatic transformations of the whole society determined by the introduction of the microprocessors, in our oppinion, the whole domain of visual arts will be completely modified by computers in the following twenty years.

Those artists or art specialists who will not be able to cope with computers in their domains of activity will soon remain behind. Some considerations about the possible influences of computers in visual arts for the following decade are presented. Some ideas for reducing the gap between East and West in the introduction of computers in visual arts are suggested.

BIBLIOGRAPHY

Kopacek, P. (1987). Education in Robotics. Preprints of the 10th World Congres on Automatic Control. Vol. 5, p. 279-284, July, 1987, Munich, Germany.

Franke, H.W. (1971). Computer and Visual Art. Leonardo, Vol. 4, p. 331-338.

Tijus, C.A. (1988). Cognitive Processes in Artistic Creation: Toward the Realization of a Creative Machine. Leonardo, Vol. 21, No. 2, p. 167-172.

Cornock, S., Edmonds, E. (1973). The Creative Process where the Artist is Amplified or Superseded by Computer. Leonardo, Vol. 6, p. 11-16.

Sirbu, D., Kopacek, P. (1991). Computer Aided Creative Process in Fine Arts. Proceedings of the First IFAC/IMACS/IFIP Workshop on "Cultural Aspects of Automation CAA'91". p. 67-68, October , 1991, Krems, Austria.

Franke, H.W. (1985). The New Visual Age: The Influence of Computer Graphics on Art and Society. Leonardo, Vol. 18, No. 2, p. 105-107.

Cultural Barriers for Work Automation in Developing Countries

E. Oliva-López and F. Bojórquez
Professional Interdisciplinary Unity of Engineering Social and Management Sciences
Iztacalco D.F., México

ABSTRACT

Automation of some critical tasks in badly needed in developing countries, in order to achieve a minimum welfare level of the population and for reaching global competitiveness in key industrial sectors. This work examines the main barriers impairing work automation in the Mexican case.

I. Background

Until two decades ago, work automation used to be concerned with the transfering of industrial tasks from men to machines; however, nowdays it is well acknowledged that many services are provided to users with the help of digital computers which have now taken under their mandate a number of intelectual tasks.

There is no doubt that industrial automation, when applied sensibly, can deliver man from dangerous, physically demanding, degrading and monotonous work. Also, as man himself takes machinery towards higher development stages, the latter becomes increasingly capable of accomplishing, in a better manner, a growing number of human tasks; thus making possible the production of more and better goods/services in an increasingly competitive way.

All this is common knowledge for people well acquainted with work automation and its potential benefits. However, there is a large proportion of the population, especially in developing countries, which knows very little about this subject and its implications for society.

2. Barriers

In the author's experience, the main barriers limiting the implementation of automation in Mexico, are:

* social acceptability
* managerial background and approach
* language constraints

2.1 Social acceptability

Social acceptability is probably the most difficult to change favourably, because it would require a new perspective of the whole community; whereas managerial background and approach and language constraints mostly depend on those members of the community which are involved in the planning, design, implementation and exploitation of automated systems.

Social acceptability is, in turn, largely determined by the existence of one or more of the following causes:

* fears of unemploment and loss of status
* lack of understanding about the meaning, implications and possible consequences of work automation
* poor technological culture
* machine phobia

As it happens in many cases of ignorance, the individual is both afraid and suspicious of something that he/she does not know. To make matters worse, many misconceptions are often created by the communications media through careless dissemination of irresponsible, biased, far fetched and commercially oriented information. All this points to the fact that, it is the lack of information about the meaning and impacts of work automation, which constitutes the main obstacle for its wider application and for a more substantial reaping of its potential benfits.

Higher levels of unemployment and loss of status are amongst the commonest of fears which people have when facing work automation. However, they mostly appear when the implementation rate exceeds the joint evolution rate of the sstem, for it is under this situation when people who have not developed themselves enough to fit in the new work structure, are irremediably exposed to the dole. This could also entail an additional evil for the organization, namel: it will have to bring in some required expertise from outside or get into its new venture with deficiencies which might become critical afterwards, at advanced stages of the implementation. Summing up, fears of unemployment and loss of status can be traced back to one or more of the following aspects:

* inherited habits, beliefs, customs and expectations
* prejudices, misconceptions and ignorance
* sceptical, uninquisitive and self-indulgent attitudes

Inherited habits, beliefs, customs and expectations are probably the main factors influency the response of a given society to new ideas and things which, in their view, threaten their current way of life. It is usually acknowledged, as the old saying goes, that: people prefer the known evil than the good to be known.

An examination of the previous list soon makes evident that:

* habits, beliefs, customs and expectations are deeply rooted into our way of living and are not easy to change
* prejudices do not follow a logical pattern and are very difficult to counteract
* misconceptions and ignorance can be considerabl reduced through suitable dissemination of information
* attitudes can be exchanged through adequate motivation

With respect to technological culture, it can be examined in two distinct and supplementary was, namely: the company's immediate environment and the community at large. In the first case, the company and other firms in the neighborhood can have an important educational impact, in the medium term, through a suitable local campaign. In the second case, only a general enlightening program can provide the required change and, even so, the expected results will appear mainly in the long term.

The machine phobia which used to be a big problem some decades ago, does not constitute a serious problem nowdays, but what is lefta of it can be treated in very much the same way that technological culture aspects.

As far as the lack of understanding about the meaning, implications and possible consequences of work automation in a given community is concerned, a community-wide familiarization program, promoted by well informed groups, should be carried out as early as possible.

2.2 Managerial background and approach

Managerial background relates to the academic instruction and practical experience which most managers have and, in Mexico, the majority of them are accountants, business administrators and other professionals with social sciences careers. Therefore, the tend to lack of the required technological basis for the suitable management of today's modern enterprises. Hopefully, there is an increasing proportion of engineers taking charge of managerial positions; however, work automation demands the full commitment of all the firm's people for success, in particular top management.

As far as current managerial approaches is concerned, the protected economic environment we were used to have, often led managers to indulge in one or more of the following sins: over confidence, laxity, negligent passivity, lack of commitment, lack of thrust, oversimplification of problems, managerial obsolescence and uninquisitive, sceptical and reprehensive attitudes.

Foresight, resolution and courage of related public administrators and compan executives, becomes essential for taking the initial steps which must precede successful work automation.

Amongst the various activities needed for a sensible application of automated systems, the most important one which should be carried out, well in advance, in order to develop the appropriate circumstances, is the development of the required infrastructure, but, people responsible for such a task is not usually aware of its crucial importance and often neglect the user's pleads about it. Therefore, this people should be the first to take notice of what work automation is all about and how it is likely to affect the welfare of all the community members.

The planning and organization sense, which community leaders should have well developed for an appropriate performing of their duties, do have in the author's opinion, a strong cultural dependence; and the rate at which such senses are developed will eventually affect the rate at which cultural change can take place.

2.3 Language constraints constitute an important barrier, because most information related to automatic systems and their components comes, for Mexicans, in foreign languages and most managers do not have a working command of any of them, not to mention supervisors and machine operators. Of course, there is alwas the possibility of having the most important documents translated, but this should only be regarded as temporary solution, for automatic systems are in continous evolution and it would be almost impossible, and certainly very expensive, to keep pace with it.

3. *Conclusions and suggestions*

As automation is now affecting ever aspect of our life, it is in every sector, activity and level of our society that wrong beliefs should be erradicated and unfounded fears should be unmasked and vanished. When supplying people people with the enlightnening information, emphasis should be put on those circumstances which greatly determine the success or failure of actual implementations.

As knowledge is enlarged from the fundamentals towards the specialized and complex, a community effort should be carried our for broadening the individual's basic knowledge and increasing his/her awareness about recent development and social trends.

It seems that the rate of change at which a given community can evolve, is naturally determined by the speed at which certain cultural changes cant ake place. This, as we saw earlier in this work, can be induced through a suitable program of actions. Technological culture is probably the single most important aspect requiring of accelerated development. Managers often refer to organizational culture as those culture-related issues concerned with the company's affairs. They have found that, through a sensible all- encompassing program, the firm can speed up the change of some cultural issues which are limiting its development. This, obviously, can only be done on those issues over which the firm can exert some control and to the extent that the outside environment makes it possible.

Good quality management in all levels of the firm, is a necessary condition for ensuring the successful implementation of work automation, for only under such leadership can most existing problems be properly foreseen and taken care of. Although this statement may appear obvious when it is mentioned, a number of unsuccessful cases have occurred because work automation has been restorted to as a substitute for managerial voids or weaknesses. Additionally, only good managers are able to use the systems approach and understand the full implications of this kind of endeavor. A good quality manager should have a satisfactory answer to the following questions, in relation with work automation: what is it?, what is its use?, when should it be applied?, where and when will it be most useful?, how should it be implemented?, who should participate and how? and what local and general infrastructure is required?

As far as language constraints is concerned, they should be tackled at three levels, namely: in the environment immediate to the company, within the company and on the whole community. The setting up of suitable policies within a given company is likely to reduce language constraints in the short and medium terms. Also, a considerable impact can be produced in the company's immediate environmet through establishing mutually beneficial agreements with all pertinent persons/organizations. Tackling these constraints in the whole community is an entirel different matter, since it implies the teaching of foreign languages as a compulsory subject of the general education curriculum. Apart from the large resources which such a goal would take and the long time needed for obtaining noticeable results, one should also realize that some members of the community could perceive a program of this kind as a threat to nationalism. On the other hand, one cannot ignore that the current globalization of the economies and the mass communications media, are already making a considerable impact in all countries, regardless of their development stage. It becomes obvious that those countries which are best prepared to follow the new global trends, will be the most benefited of all. In this context, it is essential that all community memebers understand that: "automation is not an end in itself, but it is a precious resource for man's self-realization".

BIBLIOGRAPHY

Renner, K. (1987). Strategic Management for Planning Plant Automation. Food Engr. Vol. 59, No. 1, p. 118-121.

Souissi, D. (1986). ¿Cómo preparar el terreno para la automatización? Manu. Almac., año XXI, No. 202, p. 78-79.

Wiley, D. (1986). Automation Technology: past, present and future. Prod. & Inv. Mgmt. Vol. 27, No. 4, p. 10-19.

A Professionally-Disillusioned Generation - Can There Be Found Any Remedies?

O. Pastravanu and C. Lazar
Automatic Control Department
"Gh.Asachi" Polytechnic Institute of Iasi
Iasi, Romania

ABSTRACT

The requirements in the field of control engineering in the Romanian industry of the eighties and aspects of the employment of the young graduates are presented. Social and economical consequences and perspectives are also discussed.

1. Introduction

The Romanian higher education belonged to a socio-economical life guiding centralized system in which the highly qualified unemployment problem was seemingly eliminated; all the graduates were granted jobs by the government, being compelled to work at least 3 years in the respective activity. In fact, after 1980, this mechanism of alleged social protection had as its actual goal the directioning and forced settling of young intellectuals in country side or in small towns where paradoxically there were only very few demands to meet their professional capabilities. At the same time, although there were plenty of demands for highly qualified professionals in great cultural and industrial centers, the young graduates were not allowed to be hired in those cities during the first 3 years after their graduation (a compulsory service period). Derogations were not permitted even for students of great academic merit; only after those 3 years wasted in totally inappropriate activities could they be accepted to universities and research institutes. Moreover, after 1980 the doctoral programs were practically suspended.

2. Control Engineering Departments Attracted a Significant Part of the Best High School Students

As a consequence of these socio-economical contradictions, the most frequent solution chosen when finishing high school studies was to apply to a faculty offering a professional qualification compatible only with urban life. In this respect, electrical engineering and especially electronics, control and computer engineering represented the best perspectives. So most high school students enrolled in Electrical Engineering faculties, avoiding the Mathematics and Physics faculties due to the compulsory service period as school teachers in rural area. These undesired social conditions joined to the general illusion of developing a modern and automatized industry offered the opportunity to the Control Engineering departments (set in Romania in the seventies) to gather students of great intellectual capability.

3. Control Engineering Education in Romania of the Seventies and Eighties

Control Engineering education developed rather homogeneously in 5 great universitary centers according to a unique 5 years curriculum; on graduating an engineer diploma (French model) was granted. The teaching staff were electrical engineers and in the early seventies most of them had benefited of retraining programs in Western universities with notable traditions in control engineering. Beginning with 1980, the budget allotted to higher education substantially decreased, modern equipment and technology purchase became more and more difficult, and scientific relations with the West (even publications) practically ceased. Generally speaking, as compared to other East-European countries, academic life in Romania underwent a gloomier situation.

However the outstanding automatic control journals still included certain Romanian authors, and some papers from Romania were also accepted to the most important forums in this area. Inevitably the research and teaching interests focussed on theoretical points, because the out-dated technology did not make it possible to contribute with remarkable achievements in industrial applications. Theoretical ideas and principles were mainly illustrated only by means of numerical simulation.

4. Requirements in the Field of Control Engineering in the Romanian Industry of the Eighties

While the curriculum of the Control Engineering departments in Romania had as models those of the Western universities of the late seventies (Kopacek et. al., 1984), (Desrochers and Saridis, 1984) with special subjects such as: System Identification, System Theory, Digital Control Systems, Robotics, Control System Design, etc., the level of automation in most industrial enterprises did not overpass the late sixties standard (monovariable control loops with analogue PID regulators); the only modern equipments frequently met in industry were the digitally controlled machine tools.

In Romania there were about 10 enterprises producing mainly classical automation equipment, a central institute for research in automation with branches in 15 great towns and some research institutes in adjacent fields (airships, sea-going ships, electrotechnics and power systems); their fundamental purpose was concentrated on copying without license foreign equipment, by using original solutions and autochthonous devices in order to reduce the import from Western countries.

5. The Employment of the Young Control Engineering Graduates of the Eighties

In Romania about 350 students yearly graduated Control Engineering within '80 - '85 and about 500 students, after '85; their average graduation points were among the highest records in the technical universities. But under the conditions of the compulsory service period spent in small towns offering no appropriate jobs, the great majority of the graduates could not practise control engineering, being compelled to specialize in certain related professional area (electromechanics, power electronics, design and implementation of programs for production planning and book keeping, computer service, etc.). Thus, after the compulsory 3 years service period, the great majority still prefered to maintain that kind of jobs even when they established in big cities where there were some premises for practising control engineering (see section 4). Some statistical reports show that less than 25% of the Control Engineering graduates of the eighties are engaged in specific jobs, whereas in many jobs requiring automatic control knowledge, experienced engineers specialized in electronics, electromechanics, and even technologists in those particular fields were prefered (Hanganut, 1991). The presence of Control Systems and Automation taught courses in the curricula of all technological universities and especially in Electrical Engineering education favoured the nonspecialists in the competition with the specialists in control engineering, under the condition of a poorly automatized industry (see section 4).

6. The Socio-Professional Background the Control Engineering Graduates of the Eighties Have to Face Nowadays

The retechnologization of the profitable industrial enterprises in Romania in partenership with Western corporations will represent the framework for creating many jobs for control engineers; such jobs might not be taken by nonspecialists due to the complex automatic control problems involved. The compulsory service period being annulled and the graduates of the latest years being allowed to contest for all kind of jobs, the question arises whether the Control Engineering graduates of the eighties (highly disillusioned by their previous activities) might compete with the recently trained and better professionally-directioned new generation. What percentage of the 75% Control Engineering graduates of the eighties (forced to quit their profession proper due to urgent socio-economical reasons of that time) could be reemployed in control engineering at the beginning of the nineties ? And is this readaptation really efficient? How many of the 25% being mostly employed in classical automation problems could evoluate towards the modern fields of real-time control, distributed control, robotics, CAD/CAM ?

7. Some Solutions and Perspectives

It is obvious that the individual effort and will to turn to account the present chance of practising control engineering constitute the basis of the attenuation of this socio-professional crisis. However some general solutions are foreseen by means of organizing post-university Control Engineering schools able to create the refreshing atmosphere required by the professional adaptation of the graduates of the eighties to a new technological standard. The support recently granted by the European Community to Romania, aiming at the development of our higher education system represents an important aid for bringing up to date the graduate and post-graduate taught courses (TEMPUS, 1991).

Furthermore, the involvement of Western partners, interested in industry retechnologization, in control engineering training with the view to an efficient implementation will mark an essential starting point for the improvement of the industry - education liason.

BIBLIOGRAPHY

Kopacek, P., Genser, R., Troch, I., Weinmann, A. (1984). Automatic Control Education in Austria. Preprints of the "9th IFAC World Congress", Budapest, Vol. IV, p. 112-115.

Desrochers, A. A., Saridis, G. N. (1984). Educational Trends in Automation in the United States. Preprints of the "9th IFAC World Congress", Budapest, Vol. IV, p. 116-120.

Hanganut,M. (1991). Alternatives for Control Engineering Education (in romanian). Proc. of the 6th National Symposium on "System Theory", Craiova, Vol. I, p. 174-178.

TEMPUS Vademecum (1991). Academic Year 1991/92. EC TEMPUS Office Brussels.

Comparative Study of Computer Integrated Manufacturing

C. Imamichi
Mitsubishi Electric Corporation
Tokyo, Japan

ABSTRACT

This paper discusses cultural influence on CIM system. For the first, a methodology for evaluating the influence of culture on a system is introduced. Then, this methodology is applied on computer integrated manufacturing (CIM) system in the three regions. The results suggest more consideration about the difference of cultural and social status of each country should be taken in the system design.

1.   *A Methodology for evaluating the influence of culture on a system*

In this chapter, Sugita's methodology (Sugita et. al., 1986) for evaluating the influence of culture on a system is introduced.

A new technology or a new system introduced to a society affects to the society and the technology or the system receives influence from the society at the same time. This mutual affection causes changes to the both side. This phenomenon is presented with a model: existing X receives influence from Y and results Z.

The changes that will occur in this model are grouped into the following six patterns:

1) Diminution $<Z=0>$: X invaded by Y diminishes and Y also diminishes.
2) Rejection $<Z=X>$: X rejects Y and only X remains.
3) Replacement $<Z=Y>$: X disappears and Y replaces it.
4) Coexistence $<Z=X+Y>$: Both X and Y remain without any change.
5) Combination $<Z=X:Y>$: X and Y are mixed up and complementing with
                            each others.
6) Regeneration $<Z=X\#Y>$: Characteristics of both X and Y vanish and a new
                            charasteristics appears.

The reason why such changes occure is that a new technology or a system is filterd with the measures of value brewed by the culture. Core part of the measures of value is sense of beauty and perception of analogy which are strongly related with culture unique to a society. Here culture means the whole spiritual activity but if we want to compare one culture with the another, we need some measures. Degrees of artificiality, degrees of complexity, degrees of redundancy of systems that exist in the society represent the culture.

When we want to know how a system is influenced by the culture, we should look at th system from four aspects; administrative, environmental, human and social aspects. From this view point, a system evaluation chart is proposed as a tool for comparative study of a system. (Fig.1)

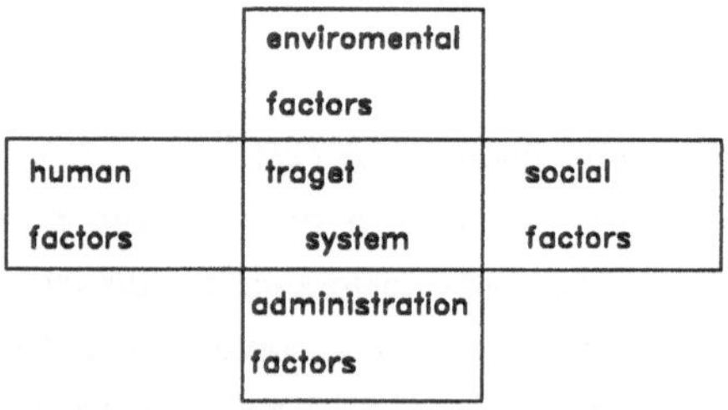

	enviromental factors	
human factors	traget system	social factors
	administration factors	

(a) Total system evaluation chart

	information	objects	human
Application	A1	A2	A3
Production	P1	P2	P3
foundation	F1	F2	F3

(b) Evaluation matrix of each factors

Fig. 1 System Evaluation Chart

The system evaluation chart is composed of five blocks. The center block shows the target system. The other blocks surounding the center block are administrative, environmental, human and social factor respectively. Contents of blocks are expressed with a matrix of elements of application,foundation, production, human, information, and objects.

2. Surroundings of The Three Economic Regions

According to the Sugita's methodology, surrounding factors of CIM system in each three economic regions (North America, Europe, and Japan) are described in the following sections.

2.1 Administrative Factors

Europe is now preparing for starting the gigantic economical union (EC) by the end of 1992. EC's population will reach 32.4 million and nominal GDP will exceed 397.7 billion ECU. EC will become comparable with U.S. in GDP and exceed U.S. in population. EC administrations concentrate efforts on eliminating physical, technological, and tax obstacles.

On the other hand, the U.S. has a plan to create a North American free trade pact among the U.S., Mexico, and Canada. This three nations pact aims to create the world's largest tariff-free trade block, with 360 million consumers and annual output worth almost $6 trillion.

The urgent problem Japanese industries' confronting with, is short of labor. Small enterprises are endangered by short of young and middle aged workers today. This trend will be much enhanced in the succeeding ten years. Japanese government estimates, it will become short of 2.5 million male workers, aged 30 to 44, in the year 2000. This figure suggests that Japanese industries will lose their power gradually and the increase of incoming foreign workers will be inevitable. Administration should prepare legal system for it.

2.2 Environmental Factors

Development of industrial society brought variety of ecological problems like as deforestation, air polution, acid rain fall and nuclear pollution. Industrialization accompanies increase of energy consumption, material consumption and discarded materials. How to save energy, is a serious problem in the situation to select nuclear power or fossil power. Europe depends their power source heavily on nuclear. The U.S. and Japan depend on nuclear less than Europe.

Technological environment should be considered, too. Core technologies to construct CIM system are electronics equipments, such as programmable controllers, robots, computers, data networks, and information processing. If a country depends such key technologies on imports, there may be difficulties in maintenance or training to use them.

Needs for electronics equipments and supply in each region are shown in the literature (Hirota et. al.,1990). Table 1. shows it's excerpt.

	Region	Needs(billion $)	Domestic Supply(billion $)
Computers			
	Europe	41.9	31.9
	Japan	30.4	39.7
	United States	56.3	58.2
Other			
Electronics	Europe	5.2	5.4
Equipments	Japan	4.5	5.6
	United States	7.6	7.9

Table 1. Needs and Supply for Electronics Equipments (1988)

2.3 Human Factors

Table 2. shows the recent trends of robot installation (Hirota et.al.,1990).

Country	1981	1985	1988	G.R./Year
U.S.	6,000	20,000	32,600	27.4 %
Germany	2,300	8,800	17,700	33.8 %
Italy	450	4,000	8,300	51.6 %
France	790	4,150	8,026	39.3 %
U.K.	713	3,208	5,034	32.2 %
Sweden	1,125	2,046	3,042	15.3%
Japan	21,000	93,000	175,000	35.4 %

Table 2. Numbers of Robots Installed in U.S.,Europe, and Japan

In the past, automation replaced human workers and, for that reason, sometimes they opposed to introduce robots or any other automatic machines into their workshops. Large scale automation made workers feel anxious because it is difficult to understand whole mechanism of controls. Daily work that need slight muscle strength make workers lose sense of real job.

Thus it becomes a problem that how human workers can co-exist with automatic machines and keep their position and safety while they work in highly mechanized process. Numbers of robots installed become the symbol of human-machine co-existence.

2.4 Social Factors

The CIM system influence on both social mechanism and fundamentals. For example, shortened products life cycle enhances the development of some new products and that may cause recycle problem. Or, bar code system becomes not enough for variety of consumer goods and retailers are forced to review the system. So many phenomenon can be seen when we check the relations between CIM and society. But the most fundamental factors which support the CIM system may be education. Industrialized countries have been making efforts to promote computer literacy education (Nishinosono,1986).

(1) North America

The U.S. is carrying out curriculum through elementary school to high school that contains the following items: Roles of computers in human life,
histories of computer development, structure of a computer, skill to handle key board, programming, use of application functions, explanation of computer related jobs

(2) Europe

France, U.K. and other European countries are also trying to enhance their education for computers. France, for example, has curriculum that covers from elementary school to high school. It contains following items: History of information system development and its effect on the society, technology of computers, programmable controllers, and robots, software development,
problem analysis and solution by computers.

(3) Japan

Japan started to prepare computer education curriculum since 1985.
In elementary school, it aims to give chances for children to touch computers and to become familiar with them. In junior high school, computers are used to do simulation or data-base look up in teaching, but independent course have not been planned. In high school, optional courses are prepared.

3. Comparative Study of CIM

As mentioned in chapter 2, each region has different surrounding factors resulted from different cultural activity. So, there is a good reason to guess that CIM systems have different characteristics for each different region. If there are differences, they show the influence of culture.

3.1 Methodology of the comparison

In this chapter, CIM systems of the three regions are compared with each others. But it is not easy to grasp CIM system objectively. I chose the following six items as for the index to show the characteristics of a CIM system.

1) Main purpose of the system
2) Scope of the system functions
3) User of the system
4) Human role in the system

5) Key issue of technology
6) Obstacles of the system construction

From these view points, I have compared 23 papers (5 from U.S. and Canada, 7 from Europe, 11 from Japan) which have been randomly selected.

3.2 Results of Comparison

3.2.1) Main purpose of the system

It was pointed out that there are two types of CIM reference model (William, 1989). The one is a complete enterprise model including management and external influences, and the other is a model confined to the factory itself. This difference is reflected on the purpose of the system. Europe papers rather emphasize the inside of factories and Japanese papers emphasize the relation with sales.

(North America)
- to improve the overall process and business operation (William, 1990)
- to optimize productivity of the manufacturing process within the enterprise (Cloutier, 1988)

(Europe)
- to improve enterprise competitiveness, adaptability and flexibility of operation/organization and to use enterprise asset effectively (Lavie, 1988)
- to increase market share of specified product range by targeting at a certain level of price, quality , terms of delivery and commitments on long-term maintenance / spareparts supply (Weston et. al.,1989)

(Japan)
- reducing lead-time, quick gathering of market needs, cost reduction, matching of sales plan and production plan (Nakano et. al.,1990)
- direct connection of sales and production function, reduction of total lead-time (Kawasaki et. al.,1990)

3.2.2) Scope of the system functions

Scope of function is decided from the purpose of the system. Accordingly the widest scope covers total enterprize activity.

(North America)
- corporate management, finance, marketing and service, research and development, production design and production engineering, production management, procurment, shipping, wastematerial treatment, resource management, maintenance management, shopfloor production (Graffe, 1989)

(Europe)
- production management, manufacturing pre-planning, computer-aided design, manufacturing process planning, manufacturing engineering of raw materials, design of raw materials, fixture design, tool design (Szelke, 1990)

(Japan)
- corporate management, accounting, sales and marketting, research and development, design and engineering, production and operation (Tachi, 1990)

3.2.3) User of the system

Users of the system are not so clearly defined. Japanese papers assume wider user of the system.

(North America)
- head-office and factory (Cloutier, 1988)

(Japan)
- headquarter and factories, domestic and overseas subsidiary, branch offices (Kawasaki, 1990)

3.2.4) Human role in the system

How human workers co-exist with highly automated system, is a big problem in industrialized society. It was pointed out that high technology induces distortion in social structure, workers' circumstances, technology inheritance, and spiritual life (Mori, 1990). But this problem was not discussed in most of the papers.

(North America)
- to make complex decisions (Jones, 1989)

(Japan)
- linked with small circle activity and visible management (Hayashi, 1990)

3.2.5) Key issue of technology

Structuring a logical reference model/implementation model, or implementaion of a system are the categories of the most papers. European papers are strongly pointing to the standardization of the reference model. On the other hand Japanease papers show not so deep concern on the standardization. They show concern on communication standards. U.S. papers pointed out practical problems.

(North America)
- data independency from system and technology (David, 1988)

(Europe)
- architectural platform of CIM system, system model, MAP (Moss, 1988)
- entity relationship analysis, petri-net analysis, client-server model, translating names and qualifiers (Busby, 1989)

(Japan)
- tracking system by magnetic card, engineering data base (Nagai, 1990)

3.2.6) Obstacles of the system construction

The obstacles common to the three regions are multivendor connection and lack of consensus what system to build. Only one paper points out cultural reason. Several Japanese papers pointed out the short of system engineers but this did not appear in other regions.

(North America)
- Complex human interface in large system, variety of OS and programming languages, unique charasteristics of multi-vendor equipment, multiple-source data, timely data aquisition, cultural resistance to share "my personal data" (Malas, 1988)

(Europe)
- technological and cost problems associated with multi-vendor information, the requirement for factory systems to change reasonably slowly area by area, uncertainty as to what will work and what will be an expensive failure, disagreement as to what basis to start (Weston, 1989)

(Japan)
- unclear concept of CIM, short of system engineer and industrial engineer, connection of multivendor network (Ohonuma, 1990)

Reviewing the result of comparison, I found several small differences but not found significant differences among the systems caused by culture. Part of that reason originate in the fact that most of the actual experiences referred is still remain in the trial level.

4.   Conclusions

I have introduced Sugita's methodology to evaluate systems in relation with culture, and applied it to CIM systems. Samples of CIM systems were got from 23 papers randomly gathered from the three regions. As shown in chapter 3, there is not significant regional differences among the CIM systems in spite of the difference in surrounding factors. That means CIM still remains in pre-fusion status (the mutual reactions stated in chapter 1 still not occured). To develop harmonious CIM systems with a society, the following points should be considered.

1) CIM is a proprietary system, but it should be considered in relation with human, social, administrative factors and environmental factors.

2) The above mentioned four factors should be taken into consideration in the phase of requirement (and purpose) definition before implementation of a system.

3) Some mechanism or methodology to derive the purpose of the systems from surrounding factors should be considered. At the same time, what are the indispensable factors that should be considered for a society when constructing CIM systems, must be agreed in the society.

BIBLIOGRAPHY

Anyouji, A., Kuwabara,H.(1990). Control Apparatus for CIM, System, Control and Information, Vol. 34, No. 3, p. 148-153, Japan.

Boelzing, D. (1989). Use of Computer-based Factory Automation in West German Companies. Int. J. Computer Integrated Manufacturing, Vol. 3, No. 2, p. 112-120, U. S.

Busby, J.S. et. al.(1990). Linking Computer Systems in a Manufacturing Company. Int. J. Computer Integrated Manufacturing, Vol. 3, No. 2, p.73-83,U.S.

Cloutier, E.J. (1988). Top Management Experience in Appliying CIM to the Enterprize. Enterprize Conference Proceedings.

Fukuda, K., Ishiguro, M. (1990). CIM in Walk-in coolers & Freezers plant. System, Control and Information, Vol. 34, No. 3, p. 178-180.

Graffe, U., Thomson, V. (1989). A Reference Model for Production Control. Int. J. Computer Integrated Manufacturing, Vol. 2, No.2, p. 86-93 .

Hayasi, M. (1990). CIM in Fuji Electric Corporation, -User Make the Production Plan by Themselves- . System, Control and Information, Vol. 34, No. 3, p. 174-175.

Hirota, K., et. al. (1990). JEIDA report on long period forecast of electronics industries in Japan, 90-91, Japan Electronics Industries Association, Japan.

Honbayasi, K. (1990). CIM and Factory Operation. System, Control and Information, Vol. 34, No. 3, p. 128-134, Japan.

Jones, A. et. al. (1989). Issues in the Design and Implementation of a System Architechture for Computer Integrated Manufacturing. Int. J. Computer Integrated Manufacturing, Vol. 2,. No. 2, p. 65-76 .

Kawasaki, O., Syouji, K. (1990). CIM in Oume Factory of Toshiba Corporation. System, Control and Information, Vol. 34, No. 3, p. 166-168, Japan.

Kosanke, K. (1990). CIM-OSA : Its Role in Manufacturing Control. Proceedings of 11th IFAC World Congress, Vol. 10., U.S.

Lavie, R. (1988). The CIM Integrated Data Processing Environment in the European Open Systems Architecture CIM-OSA. Enterprize Conference Proceedings, U.S.

Malas, D. W. (1988). Integrating Information Flow in a Discrete Manufacturing Enterprize. Enteprize Conference Proceedings. U.S.

Mori, K. (1990). High-tech Society and Labor. Iwanami, Japan

Moss, S. P. (1989). A Management and Control Architecture for Factory-floor System: from Concept to Reality. Int. J. Computer Integrated Manufacturing, Vol. 2, No. 2, p. 106-113.

Nagai, K. (1990). CIM in Oki Electric Corporation. System, Control and Information, Vol. 34, No. 3, p. 172-173, Japan.

Nakano, N., Inamoto, A. (1990). CIM in Mitubishi Electric Corporation. System, Control and Information, Vol. 34, No. 3, p. 169-171, Japan.

Nishinosono, H. (1986). On Computer Literacy Education. Information Processing. Vol. 27, No.10, p. 1129-1136, Japan.

Ohonuma, K. (1990). Data Base for CIM. System, Control and Information. Vol. 34, No. 3, p. 154-161, Japan.

Sakamoto, C. (1990). Software repertory for CIM. System, Control and Information, Vol. 34, No. 3, p. 140-147, Japan.

Sugita, S., et. al. (1986). JEIDA System Planning Working Group Report-Study of System Evaluation Chart- . Japan Electronics Industries Associatin, Japan.

Szelke, E. (1990). Intelligent and Adaptive Control of FMS Contributing to CIM Trend . Proceedings of 11th IFAC World Congress, Vol.9 .

Tachi, H. (1990). Communication Technology for CIM. System, Control and Information, Vol. 34, No. 3, p. 135-139, Japan.

Tamura, J., Matsuo, T. (1990). CIM in semiconductor plant. System, Control and Information, Vol. 34, No. 3, p. 176-177, Japan.

Weston, R. H. et. al. (1989). Configuration Method and Tools for Manufacturing System Integration. Int. J. Computer Integrated Manufacturing, Vol. 2, No. 2, p. 77-85.

Williams, T. J. (1990). A Reference Model for Computer Integrated Manufacturing from the View Point of Industrial Automation. Proceedings of 11th IFAC World Congress, Vol. 10.

Development Trends of a Culture of Safety in Computer Integrated Manufacturing and Process Industries

H.-J. Weißbach
Faculty of Economic and Social Sciences
University of Dortmund
Dortmund, Germany

1. *Expert-based and shop-floor based security cultures*

The results of our research project "Security of network-based Systems" which has run since 1990 at Dortmund University on behalf of the Projektträger Arbeit und Technik, Bonn, may illustrate that the considerations of Douglas and Wildavsky (1982) referring to high-risk systems are applicable to all types of industrial work. According to Douglas and Wildavsky, different ways of risk-perception and risk-management are not only caused by individual styles of perception or by individual competences (as research on the role of human errors in accidents sometimes suggests), but that they are also depending on cultural conditions. Security cultures as we know them e.g. from the mining industry have always included both elements of personal safety and of system security. They have been shaped, as other elements of working cultures, under the influence of technical and economic constraints of the particular industry. The results of these processes have been institutional patterns, types of knowledge, homogeneous and obligatory values, standards, technical rules, and habits of mind of professional groups (habit as defined by Bourdieu 1987, 98: as a "modus operandi") regulating risk-perception as well as co-operation and communication in risk systems.

In the course of our project, we made an empirical attempt to classify security cultures in industry on various dimensions (Möll and Weißbach, 1991). Cultural styles of risk-perception and risk-management which we call security cultures can be characterized as either shop-floor-based or professionally orientated, as individualist or collective, as group-orientated or hierarchic, etc. Collective security cultures have a strong obligatory value component. Individualist security cultures leave a broad range of decisions to autonomous operators who are responsible for their machines. If security cultures are mainly professionally orientated, they tend to develop sophisticated probability risk calculations. If they are shop-floor based, they develop simple binary codes or "patterns" discriminating situations whether they are risky or not under given technical and organizational circumstances (Tab. 1).

Safety culture	craftmen/ shop–floor– orientated	professional/ expert– orientated
individualistic	engineering	air traffic
collectivist	mining steel industry	chemical industry nuclear industry

hospital

sea navigation

Table 1. Classification of Security Cultures

In the case of expert cultures, risk calculations tend to support centralized expert decisions on the question whether the risks of complex systems are bearable or not by the operators and/or with regard to the environment (Wynne, 1988). In the case of shop-floor based cultures, the binary code or "patterns" of perception enable individuals to orientate in routine or risky environments while maintaining their individual disposing capacity.

It should be added that all types of security cultures and any form of institutionalization of risk-perception, even if they have been well-functioning for a long period, may produce particular risks (Tacke and Borchers, 1992): Whenever experts who are not personally affected by the risks which they have to calculate make the decisions, the links between risk calculations and daily experience of the operators and the communication lines between experts and operators become more and more unreliable. Even in the case of strong shop-floor based cultures with strong common and obligatory values, the operators may get insensitive to risks because of the permanent emphasis on risk protection and avoidance, or because of false inferences from many successfully passed risk situations. Some studies show that particularly elder and experienced skilled workers are endangered by disregarding and neglecting security standards after they have been often in charge of risky operations. Younger and inexperienced workers, however, suffer from an unsufficient risk cognition which leads them to underestimate distant effects rather than to neglect security standards deliberately or for pure reasons of comfort (Henter, a.o.).

2. Traditional security cultures in disintegration

Perhaps the best example for a well-functioning, highly integrated, and consistent security culture without internal contradictions is the culture of electrical engineering.

This culture, although it is rather individualist because of the isolated working situation of the electricians has formed the habits and perceptions of many people, from professionals and skilled workers down to laymen and even to children who are told to avoid touching electric sockets as soon as they leave the cradle. The culture of electric security is professionally based and brought up to date by professional asociations, but its values and standards are highly accepted and well-known on the shop-floor as well, and at least some of them are present in every household (Hölscher, 1988).

We suppose that the analysis of Perrow (1987) concerning the risks of complexity of technical systems is suffering from a certain neglect of cultural patterns. One could imagine that, with regard to the German industry, a classification of technological and organizational risks would perhaps look quite different than Perrow's classification which obviously is based on experiences of the American industry. The German chemical industry with a 100 years old security culture would perhaps protest if it is put in the same drawer as high-risk systems like nuclear plants, or only in the same as the American chemical industry.

But our aim is not to compare the risk perception in various industries in a cross-cultural and cross-national design. Rather we want to proceed to another important empirical issue of our project, namely to the question of intercultural relations and cross-professional communication in the process of development of computer integrated and process controlled production. The traditional elements of security cultures and their internal consistency are, as far as we see, highly affected by processes of automatization and informatization, and particularly by the spread of multi-component control systems and networks which are developed and implemented by specialists from various firms, professions and cultures.

As far as to the mechanical engineering industry, we observe that the traditional pattern of individual security responsability of the operator, foreman, maintenance worker, technical designer etc. for a single area (e.g. a single machine, a group of machines) or a group of components (e.g. hydraulics, electronic control systems) is no longer functional because of the domination of multi-component equipment. An overlapping security responsibility, however, does not yet exist, because the multidisciplinary ad-hoc projects of the implementation phase of new technologies are no solid structure for the following phases of adaptation, qualification, and for the collection and evaluation of the experiences with the new technologies.

In the chemical industry, we find already mixed groups of responsibles not only in the implementation phase but also in the routine running phase. There is not only the outstanding type of the academic chemist who usually had been the works manager of a production line, but there also process engineers, electrical and electronic engineers, and software-technicians. We can observe, however, that the competences for personal safety and technical security are more and more separated institutionally: The first is left to the labor directorate, the other is claimed by mixed technical technical staffs. The results of this structure are complex interactions between these professions which seem to function well, but we were aware of a new type of accident due to the fact that the technical staffs have no longer precise ideas of the vast quantity of possible compositions of chemicals and their unplanned reactions as the taditional academic chemnist as the responsible works manager had before.

A critical point of over-specialisation and multi-professionality in nearly every industry ist the unsufficient co-operation of producers and users of equipment with regard to the security problem. Big equipment ist composed of thousands of mechanical, electric, hydraulic etc. parts and controlled by software-programs which are designed by professions and produced by firms who have perhaps no idea of the future application and of the environment constraints to which it is exposed. The producers are not able to confirm that all components will work under all circumstances, and the users are not able to examine the applicability of each single component which they have bought. Moreover, the terms and classifications by which the security characteristics of the components can be described differ completely from one profession to the other. Particularly they are disregarding the context of users (as e.g. the German VDE standards do). It is nearly impossible to substitute mechanical or hydraulic by electronic components while hoping to maintain their security characteristics. It may be even impossible to translate security classifications and characteristics from one expert language to another. Some experts (Tyre, 1991) believe that most of the faults and defects of technical equipment are due to reasons which are well-known to the producers but not communicated with the users - mainly for reasons of cultural barriers, as we suppose.

In the more harmless case, traditional institutions and security rules are only dealing with the remaining trivial accidents in high-tech environments like the tumbles from the ladder. In the worse case, they have become critical to the implementation constraints and security needs of CIM or process-controlled systems. Although we agree to Ruppert (1991, 82) that the impact of the technical development reduces the time which operators and maintenance workers spend in risk areas, we draw from this the conclusion that risks can no more be analyzed at the level of single working activities or positions. The understanding of risk emergence and the need for the development of new techniques of risk management require a more dynamic and communicative concept of security. We must understand the different ways and barriers of professional perception of risks and the cultural obstacles of an open dicussion of failures which still is limited by professional and linguistic demarcation lines. It is not quite clear at the moment whether this culture will be a culture of inter-professional trust or rather of distrust (or "doubt", see Tacke and Borchers, 1992). And we are not convinced that the European efforts of standardization and of establishing common security classifications (Hölscher, 1989) will neutralize the need for reasonable doubt which the professions must have about each other.

3.   Outlines of a new security culture

Can a "culture of doubt" actually exist and play an important role in risk-prevention? Can it be implented systematically, as Borchers and Tacke propose? Can doubt be institutionalized by a hierarchic system als practice alarms can be? And who can be the supporters and upholders of such a culture?

The professional group of software engineers have got a key-role in risk-prevention in the mechanical engineering industry. Their habits and perceptions have become more and more influential with regard to security and risk-management standards (e.g. in the culture of software projects). But they tend to come into conflict with traditional habits of machine engineers, operators, and work schedulers. On one side, they are not adjusted to get along with the demarcation lines und traditional borders of transaction on the shop-floor although success in implementation projects depends highly on cross-border communication. On the other side, they often seem to be blind to the huge energy potentials on the shop-floor which are controlled by their programs, and to traditional types of risks like accidents, working-load, and health risks of the operators. But we suppose that the conflicts between the professional elements of the security culture of the engineering industry are rather due to a lack of institutionalization on the firm level than to a lack of skills or to personal ignorance.

The chemical industry tries to conserve their strongly institutionalized systems of collective security inspite of all turbulences. Although a lot of security-related institutions like the overall responsability of the academic chemnist as the works manager or the daily tour of the foreman have become obsolete, the traditional security culture and the new culture of firm security have a chance to come into a synthesis which is not any longer dominated by values and experiences of one single profession.

New security cultures, that is the conclusion from our project, have to be systematically established as firm-based cultures in processes of organizational development, and although expert involvement in these processes will be higher than it was in the shop-floor based cultures, vertical communication with the operators in both directions will be essential for the success. This includes necessarily elements of permanent doubt which can be cleared up only in institutionalized failure discourses including dialogues between producers and users.

BIBLIOGRAPHY

Bourdieu, P. (1987). Sozialer Sinn. Frankfurt/M.

Douglas, M., Wildavsky, A. (1982). Risk and Culture. Berkeley.

Henter, A., a.o. (without year): Tödliche Arbeitsunfälle. Forschungsbericht 235 der BAU Dortmund. Bremerhaven.

Hölscher, H., a.o. (1988). Sicherheitsklassen. Forschungsbericht 543 der BAU Dortmund. Bremerhaven.

Hölscher, H. (1989). Fachübergreifende Sicherheitsanforderungen - Technische Regeln im internationalen Vergleich. Technische Überwachung (30), No. 1, p. 18 etc.

Möll, G., Weißbach, H.-J. (1991). Qualifizierung in Risikosystemen. IuK-Institut Dortmund. Information & Kommunikation 8.

Perrow, C. (1987). Normale Katastrophen. Frankfurt/M.

Ruppert, F. (1991). Der Fragebogen zur Sicherheitsdiagnose (FSD). Zeitschrift für Arbeits- und Organisationspsychologie (35), No. 2, p. 77 etc.

Tacke, V., Borchers, U. (1992). Informatisierung und Risiko. In: Weißbach, H.-J., Poy, A.: Risiken der informatisierten Produktion. Dortmund (in publishing).

Tyre, M.J. (1991). Manageing the introduction of new process technology. Research Policy (20) (in publishing).

Wynne, B. (1988). Unruly technology: Practical rules, impractical discourses and public understanding. Social Studies of Science (18), p. 147 etc.

A Few Contradictory Aspects of Automation in Manufacturing Processes in the Post-Revolutionary Era in Czechoslovakia

L. Javorcik and M. Lenart
Slovak Technical University Bratislava
Faculty of Mechanical Engineering
Department of Machine Tool Design
Bratislava, CSFR

ABSTRACT

During the past eighteen months in Czechoslovakia a lot of systems have been changigng. Particularly the political system created after November 1989 brougt a new skeleton or technical and cultural policy and consequently led to the basically new approach to automation technology from some social positions.

1. Introduction

There is no doubt that automation technology has to be taken into consideration conjures up different images in many explanations. Manufacturing processes and system are comprising the products, raw materials, machines, tools, energy, know-how operations, flow of materials and information, handling processes, work organization and information management and many other complementary procedures which are also in inter-relationship with the social and cultural effects upon standard and style of people life. Under the changes in the policy system, in the East Block Countries, the evaluation aspects for automation technology have been modified and a few contradictory innovations aspects shoud be accepted by establishing the impacts in industry after the post-revolutionary era, particularly in Czechoslovakia.

This paper discusses on a few contradictory aspects which significantly effect the product development in order to reach concurrent level of manufacturing processes in machinery.

2. Has it been really so bad?

It is not unusual to hear that the Czecholovakian industry is performing the manufacturing processes on the retarded level, many technologies ar not up-to-date and automation technology has been involved into industry from the false aspects.

Looking at the position on the world market {Germany 20,6 %, Japan 18,1 %, the UK 8,3 %, ... Czechoslovakia 2,9 %} there is a reason to be feeling downcast. The German and Japanese economies have been boult on the philosophy of manufacturing expertise through industry branches which are efficient and have invested in innovating and advanced technology with the feed back into the educational system.

The current situation in Czechoslovakia has its root, of course, in the previous political system which preferred the Soviet economical model based on the rigid centralized and commanded scheme with respect to the military doctrine. This type of economy used a hard directive-administrative system practically without any private sector. The dominant approach to decision-making processes ignored or constrained any natural interest for the social or technical progress. Even politicans carry the main responsibility for the present situation. They were thinking that the political system in East Block could rach an appropriate prosperity if the "scientific" planning is working and a balance on the market will be equalized by the cunning tricks in the administrative fiel.

It is a paradox that in the UK they thought that the service branches such as finance, insurance would play a major role in their economy. In the CSFR the situation in this direction was in contrary and during the past 30 years these branches were very deeply undervalued, All this led that the products were weak in design, quality value and the range for choice. It could be argued that industry tried to complete at the non-demand Eastern markets because at about 80 % of exports were oriented to the socialist countries. Anyway, these ventures have already been behind us. A new situation, afer velvet revolution i.e. the post-revolutionary era, has been bringing dramatically changed the picture of technical policy in science, education, industry and service. At last, Czechoslovakia is going to have again the mor open economies and this all means that trade on international scale is vital to the Czechoslovak economy for the furture.

One aspect, furthermore, should be stressed. The field on Research and Development {R and D} in manufacturing processes is now becoming a dominant social factor and cheap labor has been loosing ist weight or be an argument against both the appropriate level of flexible automation coupled with the implementation of Total Quality Management based on the ISO Standards.

3. Automation vs employment?

The Czechoslovakian economy in the fifties and partially in the sixties had comparble performance to the countris such as Austria, Switzerland, the Netherlands etc. Machinery was making many successful products as in the customer field and heavy-duty industry as well, {textile, shoes, small cars, lorry-SKODA, TATRA, el. generators, machine tools, plant power turbines, locomotives, etc.}. At that time the research and development utilized the skill of workers, the inventions of the researchers as well-known the BATA management or the SKODA system. The higher educational system had many renewed professors and a good theoretical basement.

At the end of the eighties they found that the development in manufacturing processes was backward from 15 to 20 years comparing to the industrially developed countries.

The results in automation did not bring the higher production rate and the efficiency such required. In order to avoid any unemployment in industry, politicians ordered an extensive development without respect ot the sufficient of resources and environment. As for automation technology, some factories tried to install robots and manipulators, manufacturing cells or flexible manufacturing systems. A few of them, in spite of the crucial conditions,achieved the high standard in terms of quality and productivity rate. The governmental support for automation technology in manufacturing allowed to keep the prices on the national market at the relatively low level. In many cases automation technology thus was motivated by the false vision of money flowing from the state budget if a factory/manufacturer showed relevant data corresponding to the plan scheme. This criterion, of course, deformed the competitive approach for solving the problem from the economic aspect. The priority of investments was concerned to develop heavy-duty industry and so-called social employment covered over the low productivity of work. It is not secret that approximately 30 % of the disposal time for work was cut down by the idle time. One can recalculate how high volume of the production is in automation power.

4.   The Post-revolutionary era and the evaluation aspects

Today´s situation in Czechoslovakia can also be characterized by transforming processes which convert the old economic system onto a free market system. There is an increasing awareness of small and medium-sized enterprises {restructured a big factory such as ZTS Martin} in Europe due to increase in small to medium-size batch manufacture. This scope is in full coincidence with the fact that some 70 % of manufacture in the EG is now of its nature. For the next 3 or 5 years it cannot be counted with a new type of flexible manufacturing systems without some foreign investment which is being developed by industry and universities within the EG. In many cases the manufacturing companies will have to be closer towards privatization effects and international co-operation to be successful.

It ist spread opinion that the dominant motivation for automation is traditionally a need to reduce labor cost. However, our practice shows that the total automation technology is not a question from the technical aspects of a solution. It seems that social/ecological aspects play more significant role in implementation for example high-tech. Under standard conditions the benefits of automation can be far beyond a certain limit as in theory but the secondary effects must be taken into consideration very seriously if the manufacturing environment is non-steady. The first order effects such as direct labor savings can still be used for evaluation at the lower level of manufacture but the second effects such as through time, in-process inventory, material handling and consistency of quality and flexibility must be used at the higher level processes {C-tech}.

In the context said before, it is very important to support the research activities directed towards workshop oriented systems and co-operate with universities on the adoption of the ISO Standards for manufacturing base {see situation since 1992}.

There are some good starting points for design and production to create a problem oriented software useful in advanced technology. Last but not least, people mentality should also be like a relevant aspect of automation effects. To evaluate the efficiency of automation will become some clearer if one understand the other - complementary aspects such as tax policy, bank management, vast management, transporting operations etc. In Czechoslovakia all this matter will take much time but w believe that will do.

5.   Conclusion

Assuming that Czechoslovakia is to be, within the post revolutionary era, integrated into the economic system applied in the EG a few new aspects of automation in manufacturing processes should be accepted. It is not simple to determine a hierarchy and weight ratio among the aspects. If one can estimate this pattern then an algorithm of the optimal structure can be find out relatively easy. But some difficulties occur at the quantification/quotation of the aspects such as an extreme of the aim function for evaluating process. Some of non-quantifiable aspects can be expresses by expert knowledge using knowledge-elicitation techniques. To gathered knowledge then it may be generalized so that a relation needed can be easily formulated. In conjunction with expressed before the orientation on expert systems is one of strong activities for automating the manufacturing processes in Czechoslovakia.

BIBLIOGRAPHY

Javorcik, L. (1991). We are on the same floor. The TEXT general Assembly and Conference, May, 91, Bratislava, p. 6.

Javorcik, K., Lenart, M. (1988). Possibility and Problems of Artifical Intelligence in Manufacturing processes. {Ed.} Alfa Bratislava, N. 16, p. 31-40

Javorcik, L. (1991). Some Findings on Applied Research at the STU Bratislava, Report No. 01/1991, p. 10.

Philosophical Impacts on Technology

P. Kampits and E. Vogt
Department of Philosophy
University of Vienna
Vienna, Austria

ABSTRACT

In the following three postmodern positions are invoked to highlight and examine some of the structurally new features of the contemporary social space: Jean Baudrillard's theory of simulacra and simulation, Michel Foucault's concept of panoptical discourse and Jacques Derrida's strategy of deconstruction.

A transformation has been occuring in which technologies - media, cybernetic models and steering systems, computers, information processing, knowledge industries and so forth - replace industrial production and political economy as organizing principle of society. In this era, signs and codes become the primary constituents of social life, whereby the structural dimension gains autonomy, to the exclusion of the referential dimension, establishing itself on the death of the latter. The high-tech society is the site of an implosion of all boundaries, regions and distinctions between appearance and reality, subject and object, and just about every other binary opposition maintained by traditional philosophy and social theory. This technologically created situation of the simultaneity of different concepts can be best described by Postmodernism, since its ontology displays certain correspondences with the new technologies. As the latter employ interweaving, instantaneity and virtuality as basic categories, so the postmodern ontology is marked by pervasions, leaps and irritations. The convergences of postmodern ontology with teleontology are partly striking. Therefore I will invoke three postmodern positions to highlight and examine some of the structurally new features of the contemporary social space: Jean Baudrillard's theory of simulation, Michel Foucault's concept of panoptical discourse and Jacques Derrida's strategy of deconstruction.

According to Baudrillard, simulacra and simulation play such a key role in social life that previous boundaries and categories of philosophy dissolve altogether. All oppositions between appearance and reality, subject and object, the public and the private, collapse into a functionalized, integrated and self-reproducing universe of simulacra controlled by simulation models and codes (Baudrillard, 1983). Simulacra are reproductions of objects or events, while the different orders of simulacra form various stages in the relationship between simulacra and the real. Today we are in the order of simulation proper, the end result of a long historical process of simulation, in which simulation models come to constitute the world, and overtake and finally destroy representation. The digitalized technologies of communication provoke the disappearance of the local in favour of the spatial; they effect a de-localization. In the new networks of communications all positions assimilate with each other and are interchangeable with each other, insofar as every position can be reached from every other position in like manner without having a starting or a final position. Teleological processuality has come to an end. And the position of the individual within the different networks of communication makes his status as subject precarious. Moreover, without any object no subject. The position of the subject becomes untenable, since it has no localizable or materializable other to catch hold of. Models and codes thus come to constitute everyday life and modulation of the code comes to structure a system of differences and social relations in the society of simulations. Everything is reduced to a binary system whose two supposedly dominate poles cancel out their differences, and serve the maintain a self-regulating system which marks a decisive stage in the process of de-realization. Simulation creates its own realities by annihilating the distance between reality and fiction. With the words of Baudrillard: we have entered the era of hyperreality, in which the real is produced from miniaturized units, from matrices, memory banks and command models (Baudrillard, 1983, p. 10).

In association with the rise of new technologies an information industry of seizable proportions has grown up, in which the gathering of datas on individuals becomes the main goal. In the context of computer networks, every scrap of data has the possibility of multiple connections and so acquires an expanding, unpredictable number of applications. There is no limit to how much data such matching and modeling systems can absorb. Conceivably, no pool of information is too small to be sponged up by technology. All that is needed are traces of behaviour; credit card activity, traffic tickets, telephone bills, fingerprints, library records, etc. In order to make intelligible ways in which new technologies like the data base generate new - linguistic and symbolic - forms of domination, the analysis must refer to a theory framed by assumptions that are commensurate with the kind of formation that are produced in the new information machineries. Therefore I will focus on Foucault's concept of discourse as presented in "Discipline and Punish". In this text Foucault underscores the relation between written and stored knowledge on individuals and institutions, whereby the effect of examinative discourse may be discerned only if discourse is grasped as a language formation, which includes the statements possible to affirm, the rules for the formation of these statements, and a system by which such statements are validated in the disciplinary community. Discourse analysis gives interpretive priority to the language formation over the subject, reason and idea in order to uncover language pattens which, when associated with practices, position those practices in definite ways and legitimize the patterns of domination inherent in those practices. In "Discipline and Punish" a set of discourses consisting of the Enlightenment reformer's tracts advocating the abolition of torture, the records kept by the administrators of the prisons and by police, and the sciences that accompany and legitimized the prison, is subject of the operation of discourse analysis. Foucault shows how discourses and pratices are interwoven in the fabric of the prison. The crisscrossing interplay of these discursive registers mentioned above is labelled by Foucault Panopticon. The term Panopticon denotes an artifice that combined all the elements of the systems of exclusion and discipline like organization of bodies, spaces and locations, supervision and regulation. The major effect of the Panopticon is to ensure a continous and complete surveillance. The panoptic mechanism of power is automatic, anonymous, and continuous. Finally the Panopticon provides an apparatus for controlling its own mechanism. The controllers and the controlled are all part of the machine. According to Foucault, the panoptical discourse as a means of control and discipline permeates the whole society. It is imposed by the systematic generation discourses, by the continual monitoring of daily life, adjusting and readjusting ad infinitum the norm of individuality.

The same panoptical system has been perfectly and widely extended in the second half of the twentieth century by dint of the computer's ability to gather and store information. Foucault claims that surveillance by means of the information technologies is something new: "Our society is not one of spectacle, but of surveillance; under the surface of images, one invests bodies in the depth; behind the great abstraction of exchange, they are continues the meticulous, concrete training of useful forces; the circuits of communication are the supports of an accumulation and a centralization of knowledge; the play of signs defines the anchorages of power; it is not that the beautiful totality of the individual is amputated, repressed, altered by our social order, it is rather that the individual is carefully fabricated in it, according to a whole technique of forces and bodies" (Foucault, 1977, p. 217). It is now possible to monitor large population without the material apparatus of the prison. Electronic monitoring of the population occurs silently, continously, and automatically along with the transactions of daily life.

The ubiquitous imposition of the panoptical metadiscourse generates a new language situation that has unique, disturbing features. Language has to be adjusted to the possibilities of encoding of the information machinery. Whatever is not homogeneously transformable and translatable into information quanta, has to be excluded or marginalized. The postmodern philosophy is sensitive to the unification and standardization of language. It resists against this new universal idiom, which is technologically formed and economically demanded by turning against this actual hegemony of one discourse, of one language game over others it linguistic instruments. Against the annihilation of the incommensurable, against the pseudognostic dissolution of the opaque into telematic sequences of information Jacques Derrida erects his concept of writing that accomplishes a deconstruction of technologically reduced language.

Although Derrida does not problematize the question of technology in an explicit way, he always already answers to it by marking metaphysical thinking, which he links up with the concepts of sign, language and writing. He interprets the metaphysical episteme from the perspective of the structurality of structures (Derrida, 1967). The virulent polysemy of structures gets metaphysically reduced by instituting a clear, absolute center - a logos - , from which meaning and sense result. This logocentrism defines the Being of beings and the sense of Being as presence. Crucial is, that the presence of Being has always been conceived as phoné, as voice. The logocentrism is also a phonocentrism. This privilege of the voice however cannot obscure the fact, that in its function as signifier it is a technical derivative. The instrumentality and technicity of the relation between phoné and Being already infers from the continuing derivative and representative relation between phoné and writing. Writing has always been conceived of as mere mediation of a mediation, as loss of the interiority of sense and meaning. And it is this relation of representation that has dominated the Western concept of writing.

Beyond that, the question of technology is somehow implicated in the question of writing, since a certain questioning for the meaning and the origin of writing precedes a certain questioning for the meaning and the origin of technology (Derrida, 1974, pp. 45 ff.). Both concepts focus on the fundamental technicity of Western thinking. Since the logocentric metaphysics has given rise exclusively to the technical character of writing, technique and logocentric metaphysics are tied together. The logocentric metaphysics, which defines the sense of Being as presence, has emerged within a system of language, that is centered around phonetic-alphabetical writing. The history of writing has always been a technology and a history of technique interpreted as an instrument for communication, that simply conveys meaning. Derrida wants to think the other of technicity as an access to multi-dimensionality by radicalizing the concept of writing in such a manner that a reaffirmation of those features of language becomes possible that have not been noticed by logocentric metaphysics. It analysed texts as self-enclosed systems of definite, stable meaning and reference - as books -, their signifiers all pointing back towards some "transcendental signified" or source of authentic and unitary truth. The traditional idea of the book is of a writing held within bounds by the author's sovereign presence; a writing whose integrity of purpose and theme comes from its acceptance of these proper, self-regulating limits. To question the authority of the book is also to challenge the priority of speech over writing, presence over absence, the origin over that which merely repeats or inscribes the origin. Against this idea of a controllable totality the textuality of writing has to be invoked. This term could be delineated as a relational network of instituted traces in general. This network of traces is a differentiating activity that covers a field as the common nexus or frame of interchange; a field where discrete points of origin, beginning, locus and situation are irretrievably buried in the ensuing play. It is a field of diffuse activity which produces meaning by differentiating words spatially and temporarilly. What is important about this play of differences is that it defines the role of the subject in the way that the game defines the role of the player. The subject is a placeholder within certain pre-existant structures of language whose resources it is never fully aware of. Therefore a move occurs from a search for metaphysically fixed meaning to an exploration of the ambivalent play of differences in writing. This consistant disruption of logocentric discourse and its attendant subject are positioned at a transitional point in history: the program of deconstruction appears when the age of the book is over and new forms of writing/language, including electronic writing, announce a new, "monstrous" age: "The development of the practical methods of information retrival extends the possibility of the "message" vastly, to the point where it is no longer within translation of language, the transporting of a signified which could remain spoken in its integrity. It functions ... without the presence of the speaking subject ... phonetic writing ... is limited in space and time ... This conjunction of cybernetics and the "human sciences" of writing leads to a more profound reversal" (Derrida, 1976, p. 10). And as a critique of phonetic writing, deconstruction seems to be commensurable with the technologies of electronic writing insofar as its theme of writing is validated by the new forms of language. This can be confirmed by interpreting computer writing as a kind of deconstructive practice that destabilizes, subverts or complicats certain features of logocentric metaphysics. Some of its effects could be outlined the following way:

Computer writing displaces the discoursive practice of authorship by positioning the subject at a level different from the conventional author/text relation. The self is de-centered, disseminated, dispersed into electronical traces, remainders. Those put the self-decentering into practice by transgressing logocentric metaphysics and its pre-critical relation to the signified. They testify both to the impossibility of appropriation of one's ownness and to the process of ex-ap-propriation, of decentering the self (one's own). If the self is disseminated into the electronic writing, which is itself dispersed, what does one own? Is that which one writes his writing? Is it his property? Can one copyright it such that one can make claims to it? And what of one's proper name? Once again the self decentering goes to work. In electronic writing, the identification of a proper name with the self is at least complicated.

With its dispersal of the subject in non-linear spatio-temporality, its decentering of the self, computer writing instantiates the deconstructive strategy of inscribing the other into the ownly sphere of the self. One might call it a monstrosity.

BIBLIOGRAPHY

Baudrillard, J. (1983). Simulations. Semiotext(e), New York.

Derrida, J. (1976). On Grammatology. John Hopkins University Press, Baltimore.

Derrida, J. (1978). Writing and Difference. Routledge and Kegan Paul, London.

Foucault, M. (1977). Disciplin and Punish. Pantheon, New York.

The possible connection of Foucault's discourse with the new technologies was first outlined by Poster, M. (1985). Foucault, Marxism, and History. Blackwell, New York.

Automation Might Mean Search for Freedom

Elohim J. L.
Mexican Association of Systems and Cybernetics
Cuauhtémoc, D. F., México

ABSTRACT

Automation seems to be, at present, a good alternative to organize systemically, cybernetically and synergetically an intellectual revolution in the human environment: the Knowledge Revolution; it must allow the members of the human species to become aware about the relativeness of their freedom to push ahead consciously but also rationally their development.

If the mankind were to bear witness of the cultural transcendentness of the increasing automation everywhere, it would certainly face a paradojical situation derived from the presence in parallel of two kinds of facts, those that would apparently push forward the human species' development and those that, no doubt preclude such development

At present it seems proper to assume that the main concern of the homo-sapiens-sapiens is to become aware enough about the interdependence that exists de facto between the evolvement of several technological actions in every particular society and the individual and collective progress of people involved. This interdependence might be interpreted in terms of mutual cause-effect relations between the TECHNOLOGY'S EVOLVEMENT and the MANKIND'S DEVELOPMENT altered eventually by many other factors which sooner or later are necessarily identified (Fig. 1).

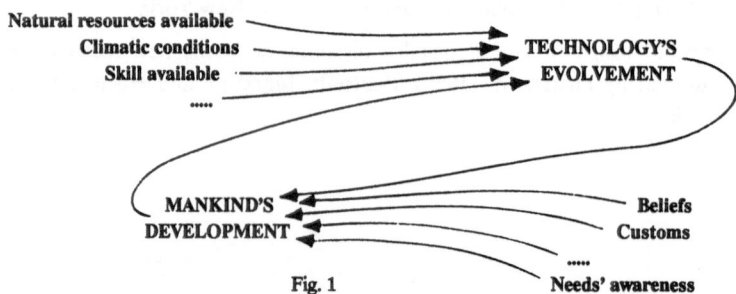

Fig. 1

Being aware of this question has become quite an indispensable task to deal with, for learning gradually how to manage rationally its manifestation, which means to search good alternative measures -accordingly to universal ethical and ecological criteria that have to be integrated step by step- in order to cope successfully with an increasing number of difficulties, which are arising always from accumulated lacks of concern about the effects of technological automation upon culture and viceversa.

The temporal character of this phenomenon makes advisable to observe it, first of all RETROSPECTIVELY, in order to get the indispensable knowledge required for seeing it PROSPECTIVELY afterwards.

Looking back to the role played by technology in diverse environments allows to recognize that it has been the main source of larger possibilities offered for organizing actions addressed to push ahead the community's progress and the development of cultural values. Both have certainly caused the building of civilization.

Looking thoroughly the causality of the technological business permits to assert that it has been essentially provoked by two fundamental activities:
* the identification of social needs,
* the assimilation of knowledge about diverse aspects of the real world (Fig. 2)

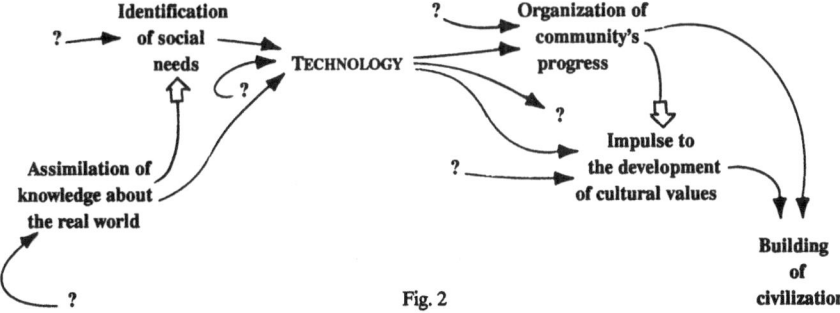

Fig. 2

Hominids first and human beings afterwards, being technologically minded, have been learning through perception of facts,... comprehension of possibilities and restrictions,... abstraction of details,... assimilation of essential aspects,... to take advantage unilaterally of an ever increasing number of objects that they find in their surroundings. Apparently they have assumed that natural things (men included) and things produced by men are basically no more than "objects" to take or to use,... resources to exploit or to transform,... and when necessary to reuse,... to throw them out or to put them aside.

One way or the other, supported essentially by this attitude, men have apparently succeeded in building peculiar artificial worlds which have been inserted gradually and deeply in the natural one, by means of cultural innovations, sometimes despite the presence of cultural bonds, some other times supported by these bonds. A "magnificent" civilization is the outcome of this cultural "mixture". How these events have taken place ? It seems that always, in accordance to particular circumstances, an increasing number of possibilities for further development have arisen from successful interactions established between diverse human needs and means for satisfying them, which were made up from possibilities that had previously arisen (Figs. 3, 4 and 5)

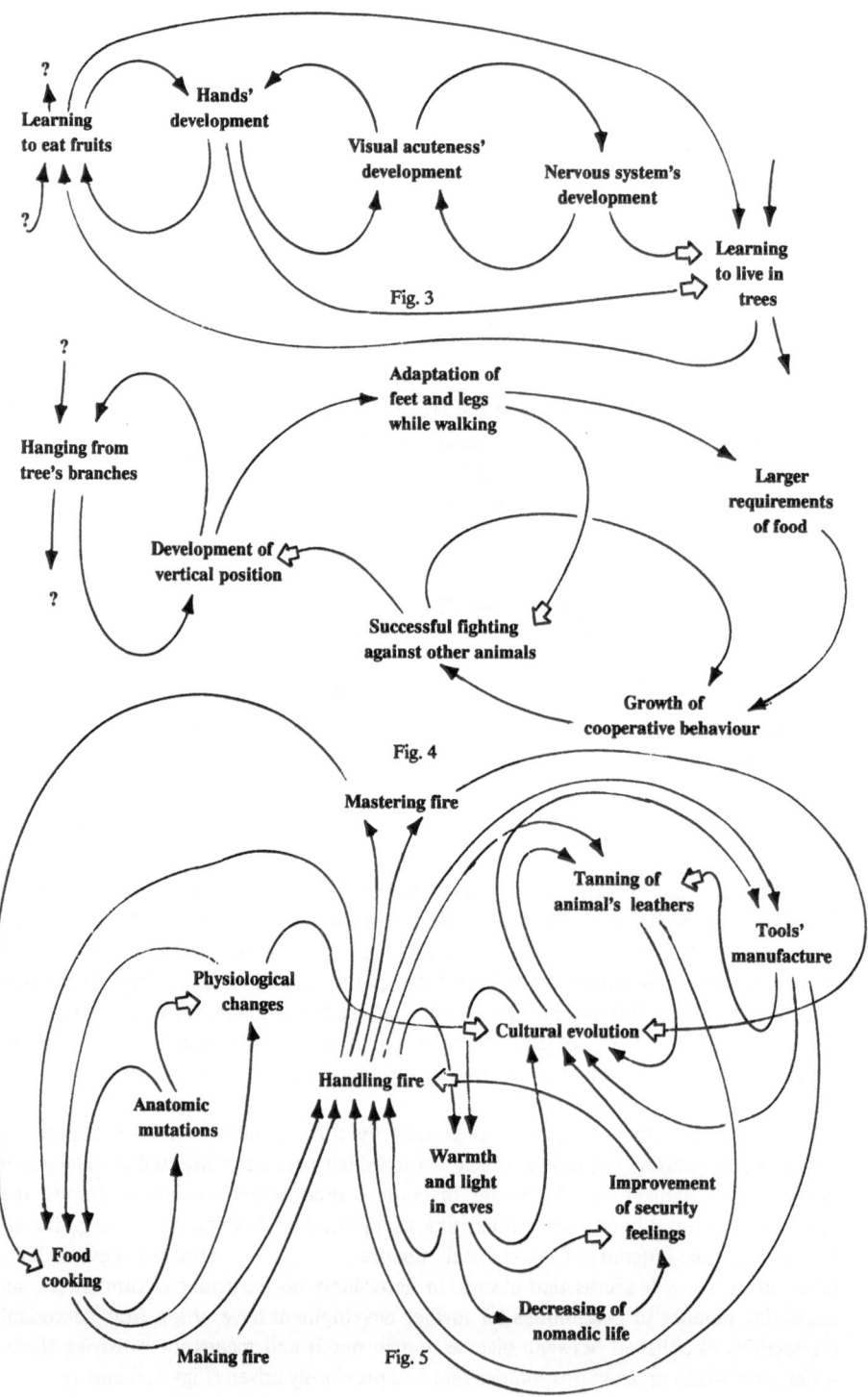

? Learning to eat fruits

Hands' development

Visual acuteness' development

Nervous system's development

Learning to live in trees

Fig. 3

Hanging from tree's branches

Adaptation of feet and legs while walking

Larger requirements of food

Development of vertical position

Successful fighting against other animals

Growth of cooperative behaviour

Fig. 4

Mastering fire

Tanning of animal's leathers

Tools' manufacture

Physiological changes

Cultural evolution

Anatomic mutations

Handling fire

Warmth and light in caves

Improvement of security feelings

Food cooking

Decreasing of nomadic life

Making fire

Fig. 5

Diverse human performances, accordingly to many different facts and conditions, during millenia, were first conceived and implemented as cultural innovations and mantained afterwards as social bonds, making possible consequently the emergence of events that have been the source of such "splenderous" civilization. These events are the relatively well known agricultural revolutions, metallurgical revolutions and industrial revolutions.

"Man makes himself" (Gordon Childe) and, to all appearances, "he" makes it as soon as "he" learns how to take advantage of his "...immense capability for culture which is dependent on his increasing educability" (Dobzhansky and Montagu)... However, it seems that so far most men are rather unconscious and quite unconcerned about the kind of man that they contribute to make, willingly or not. Very often men's performances, which couldn't not be the outcome of their particular cultures, provoke the destruction of cultural accomplishments because they assume that their actions should be based on the development of their abilities for fullfilling reasonable goals for them, i.e. those that will satisfy their immediate interests (Figs. 6 and 7).

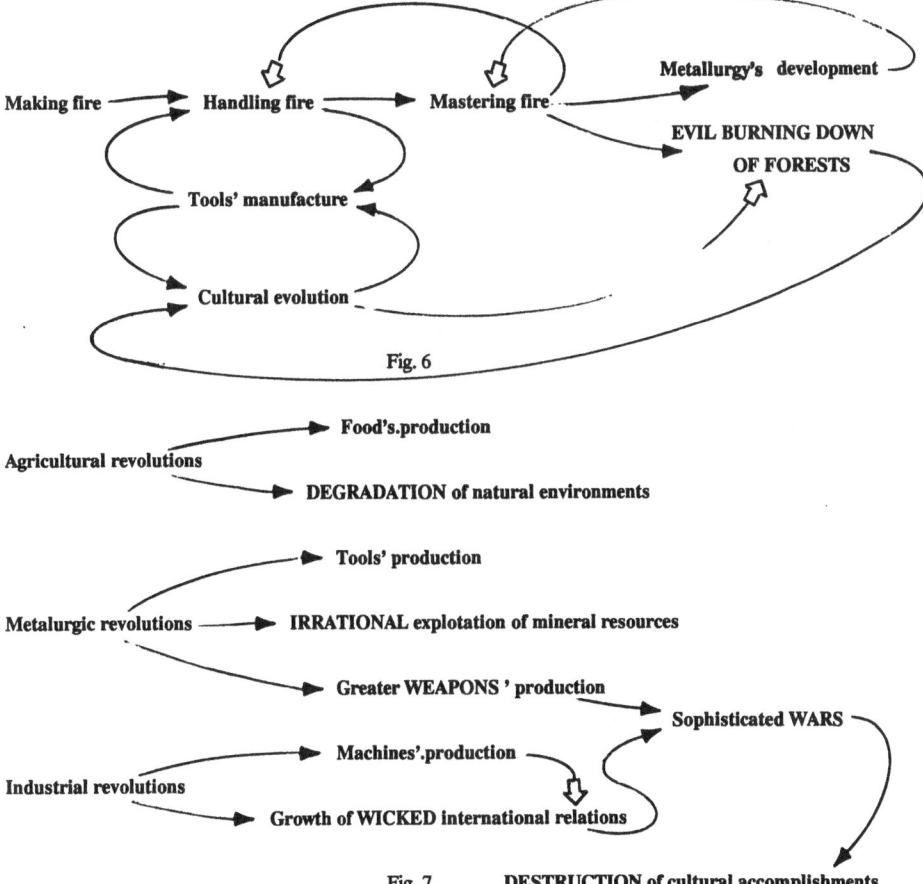

Fig. 6

Fig. 7

Why the outcome of my reasoning produces a reasonable outcome for me, while their effects are unreasonable for others ? (Fig. 8).

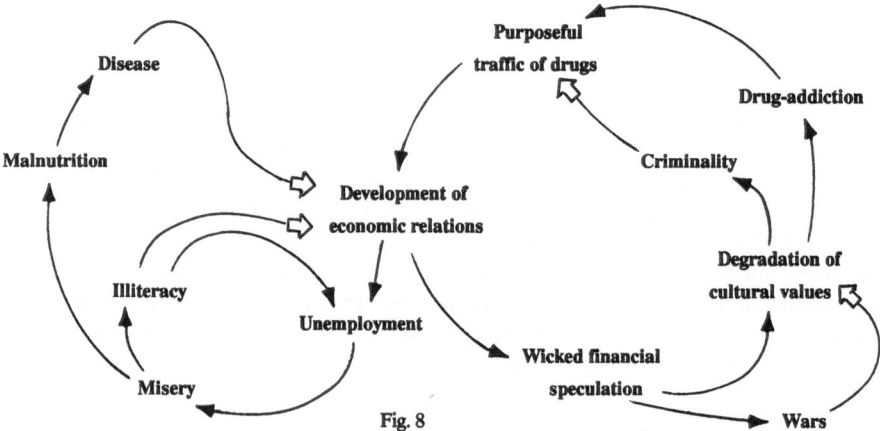

Fig. 8

A network of structured assertions (Fig. 9) seems to provide a preliminary answer.

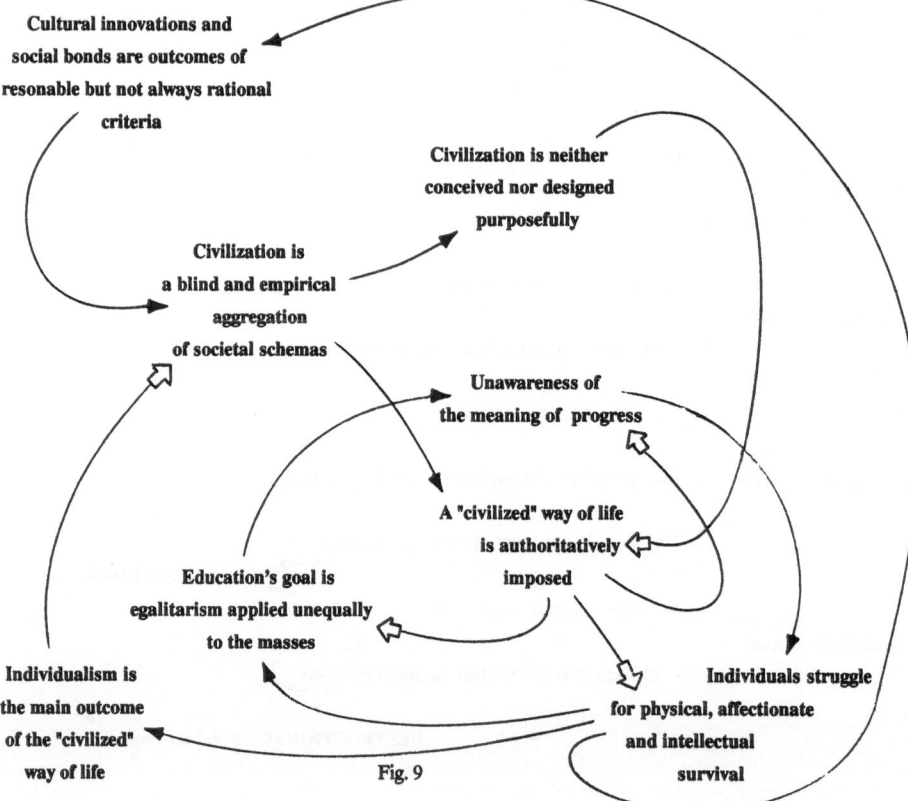

Fig. 9

In accordance with the view offered by this network it is evident that technology couldn't be responsible of the intrinsic difficulties that so far characterize the actual civilization's dynamics. However during the very recent decades the side effects of automation (certainly the most sophistcated technological manifestation) -particularly in the domains of transport, communications and data processing- clearly have hastened unavoidably the crisis of this civilization. It is so because everything in society happens nowadays faster and faster, caused by an increasing automation, while the constituents of that civilization remain quite unable to either prevent, preclude or forestall effectively and rationally the ocurrence of such events which at present provoke more and more perilous, hazardous, and risky situations in the human society. It couldn't happen other way, because this "human" civilization has been and still is very poorly related to the development of the homo-sapiens-sapiens whose members, up to this time, remain quite unconscious about the possible rationality of their "roles" on this Earth (the only one available for them, at least for the time being).

Notwithstanding this fact, technology in general and quite particularly automation have become at present an inherent potentiality of civilization for allowing their members to search out prospectively feasible alternatives for the development of the species who certainly has already developed this technology.

This potentiality arises from a cybernetic interpretation of dynamic features of social phenomena, which means to organize purposefully positive feedback actions aiming at the gradual dissapearance of some undesirable effects or the gradual growth of certain desirable consequences (Fig. 10)

Dissapearance of undesirable side-effects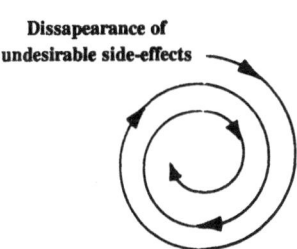

Growth of desirable consequences

Fig 10

Could processes of this kind propitiate, at least, the setting up of favourable conditions that would allow the emergence of APPROPRIATE technologies? The purposeful organization of positive feedback interactions seems to be suitable media for bringing about useful information intending to assure that the assimilation of knowledge about the real world might be ENOUGH, that the identification of social needs might be RIGHT, that the organization of the community's progress might be SENSIBLE, that the development of cultural values might be PROPERLY pushed forward, that the processes for building the pretended human civilization might be RATIONAL, that... (Fig. 11)

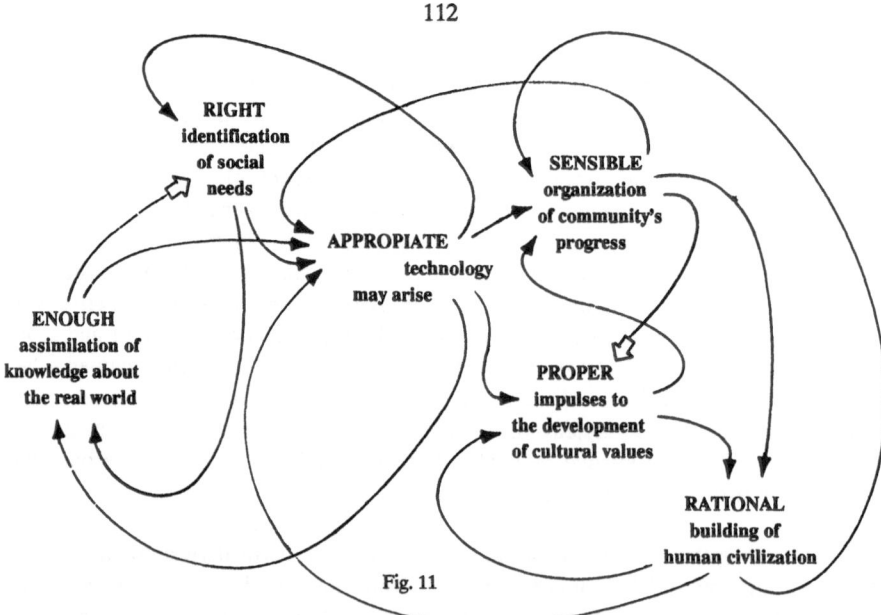

RIGHT
identification
of social
needs

SENSIBLE
organization
of community's
progress

APPROPIATE
technology
may arise

ENOUGH
assimilation of
knowledge about
the real world

PROPER
impulses to
the development
of cultural values

RATIONAL
building of
human civilization

Fig. 11

Until now automation seems to be justified: it can increase substantially either the speed or the precision or the safety or the productivity or the competitiveness or the quality or several of these production's needs; however automation risks to favour easiness instead of hard and difficult effort, simplicity instead of complexity, leisure instead of work, privilege instead of distributed obligation, passiveness instead of active involvement,...

Shouldn't automation be concerned with:
* the reduction of useless leisure time
in order to increase facilities for
indispensable work;
* the suppression of schemas for the homogenization of minds
in order to establish means for
increasing the diversity of individualities;
* the suppression of simplified approaches and methodologies
in order to support the learning for
dealing systematically with complexity and
in order to find out how to complement
apparent opposite (possible) actions;
* the reduction of trivial holidays
in order to create suitable conditions for
further development of human capabilities;
* the gradual elimination of individualism
in order to create circumstances for
becoming aware of social commitments;
* the end of empirical aggregation of societal schemas
in order to support the individuals' participation for
synergetic integration of societal systems;....

The potentiality of human mind -despite all the recent set-backs- taking into account the presence of unavoidable draw backs, intrinsic to human constitution, seems to be ready enough to conceive, design and implement on purpose

a KNOWLEDGE REVOLUTION

which should aim to allow every individual, everywhere, to become aware about the need of creating a **SYNERGETIC CIVILIZATION** through the active participation of everyone, searching out to organize at the same time the progress of his (her) community, of his (her) nation and of mankind based on two fundamental criteria:

Ethical
* To propitiate the development of diverse cultures.
* To allow the survival and even the development of every species.

Ecological
* To search out systematically a better comprehension of natural laws in the earthly environment increasingly altered by "artificial" means continuosly produced by countless technological actions.
* To push ahead consciously the development of Nature on the Earth.

May this view lead to find out the HUMAN DIMENSION of an AUTOMATION that can push forward the HUMAN SPECIES' DEVELOPMENT ?

I have not answer for this question for the time being, instead I dare suggest that the WG on Cultural Aspects on Automation of IFAC should promote the utilization of Synergetic Work in order to support everywhere actions aiming to conceive and design:
* Schemes for basic education (3 to 19 years old)
    that ought to allow everybody,
        the theoretical and practical assimilation of
            ethical and ecological principles
        in an automatized environment.
* Views to update the engineering education
    that would encourage every engineer
        to perform synergetically in accordance to
            social oriented goals and
            cultural oriented objectives.
* Outlooks to improve the panorama of social sciences
    in higher education institutions
        for allowing the social scientists
            to get updated technological backgrounds
        suitable to understand and criticize synergetically
the automation's possibilities and restrictions